I0445335

What's Eating You?

Copyright © 2026 and beyond by Doctah B Sirius

Published by

Elevation for Abundant Life LLC

This book is dedicated to my beloved sister, Tamara Branch, who has been a profound source of inspiration to me since the day she was born. She is a lighthouse, steadfast and radiant guiding not only me, but everyone fortunate enough to be touched by her light. As my greatest supporter, she pushed me to persevere, to always be my best, and to write this book. When it was finally complete, she read the first chapters and told me it was amazing, telling me how proud she was.

Just a few short weeks later, I found myself holding her hand as she took her final breath.

Tammy, I miss you more than words can express. This book is for you.

What's Eating You?

Break free from Mental, Physical, and Energetic Parasites.

The journey from opposition to liberation.

By Doctah B Sirius

This book is also dedicated to my parents, Jessie and William Branch, Courtney Jr., Anne Zene, Mr. Dick Gregory, Efa Sade, Dr. Anthony Kweku Andoh, Dr. Oba, and Chief Little Bear, who have all made their transitions.

Special thanks to Divine Intelligence, Vashti Bonner, Evelyn Metoyer, my siblings Gregory, Randy, Terrell, and Tamara B., Andrew Ravar, Larry "El" Glenn, Cierra Duval, Carl Nelson, Jamaal Gore, Richard Merrit aka Brother Rich, Dawn Osbourne, Eugene Cooke, Gregory Curtis, Ananda Lo, N.J. Johnson, Melikia Courtney, Robert Faust Ph.D., Akahdamah Jackson, Tori Reid, Dr. Mahdi Brown, Billy Carson, Dr. Michael Beckwith, Mother Kali, Cynthia Willis, Sarah Girma, Tracy Kendrick, and to my children Delaine, Justice, Aisha, Halim, Esa, and Samirah.

Special thanks to George Clinton of Parliament-Funkadelic, and a very special thanks to Iyanla Vanzant.

Advisor: Vashti Bonner
Edited by: Ananda Lo and Evelyn Metoyer
Publishing Technician/Editor: Maat em Maakheru Amen
Public Relations: Makeda Smith of Jazzmyne PR
Photography: Kawai Matthews of Air Philosophy
Cover design by: Aisha Allah

Table of Contents

Foreword

I have known DOCTAH B SIRIUS for many years, and throughout that time, I have witnessed firsthand the profound transformations he describes within these pages. His journey has been nothing short of a spiritual odyssey, one marked by courage, surrender, revelation, and a steadfast commitment to awakening. When he speaks of liberation from physical, mental, and energetic parasites, he does so as someone who has walked through darkness, emerged into the light, and returned bearing medicine for the rest of us.

Doctah B is not only a masterful teacher, but he is also a vessel of deep spiritual wisdom. His work carries the vibration of someone who has been called, refined, and guided by a higher intelligence. His teachings bridge dimensions, uniting ancient herbal science, metaphysical understanding, intuitive knowing, and the subtle energetic truths that shape human experience. Through his voice, one senses the lineage of ancestors, mystics, healers, and cosmic intelligence flowing into the present moment.

I can personally attest to the authenticity of his path. I have used several of his herbal remedies myself, and I can say without hesitation that they are effective, powerful, and crafted with a rare combination of scientific insight and spiritual discernment. His formulas do more than address the physical body; they recalibrate, realign, and restore the subtle aspects of one's being. They are living testimonies to the truth he shares in these pages.

What's Eating You? Invites us to look deeper than the surface, to recognize parasites not only as physical invaders but as energetic distortions, emotional wounds, toxic habits, and mental programs that feed on our life force. With clarity and compassion, Doctah B reveals how these influences shape our behavior, cloud our intuition, and fragment our connection to Spirit.

This book is ultimately a call to liberation. It reminds us that healing is a sacred act, that reclaiming our sovereignty is an initiation, and that we each possess the inner light capable of dissolving what seeks to diminish us. Through his story, teachings, and natural remedies, Doctah B offers a pathway back to harmony, physically, mentally, and spiritually.

I offer this forward with great love and deep reverence for a brother whose mission is to uplift humanity. May these words guide you, challenge you, and awaken the spirit of freedom already alive within you.

Peace and dynamic blessings,

Michael B Beckwith
Founder – Agape International Spiritual Community
Host – Take Back Your Mind Podcast
Author – Spiritual Liberation -Fulfilling Your Soul's Potential and Life Visioning

Preface

This book has been a long time coming, and I am grateful that its time has finally arrived. My highest choice is that this book will be used to create an abundant, healthy, wealthy, free, harmonious, and inspired life for each reader.

I am very fortunate to have had several mentors and guides. One of my most influential mentors since 1995 was the great civil rights leader, comedian, track star, natural health pioneer, educator, author, and father, Dick Gregory.

In the late 90s, he, I, and several other teachers were on a panel in Fort Lauderdale, Florida. Mr. Gregory and I sat together on the stage, and throughout the day, he pulled all kinds of rare plant medicines, healthy snacks, and natural health products out of a huge bag under the table, and for hours, he kept pulling items. And while cracking jokes in my ear, he would explain the health attributes of each, and he insisted that I take some, which I did. After a while, I felt like I was having an out-of-body experience, and a wealth of information filled my head. When I was called to speak, I couldn't remember one bit of what I'd planned to talk about. Dick Gregory patted me on the back, and I stood up in a daze; without thinking, I asked for five volunteers from the audience. I then facilitated a group levitation, demonstrating how four people can lift a 250-pound man using only two fingers from each volunteer. I guided them through a type of Subconscious Alignment process, which broke the spell of limitation as they together lifted the 250-pound man out of the chair. He was as light as a feather.

The crowd screamed in amazement. Afterward, I explained that all things are possible when you let go of limiting thoughts and learn to use your imagination properly. Then I sat down, never presenting the information I had initially prepared; I couldn't even remember it. Mr. Gregory turned to me and said, "You're definitely the one."

A short time later, Mr. Gregory demanded that I join his team of natural healers. Conventional doctors had diagnosed him with lymphoma, an incurable and fatal type of cancer. They said he only had a short time to live. He chose a natural path for his healing, and we were there to help facilitate it. He proved the doctors wrong and continued his inspirational work for many years. I was honored to work with him and was always astonished by how much he knew about parasites, plant medicines, epigenetics, quantum physics, and metaphysics. In fact, that is all we ever talked about.

Whenever I would attempt to tell him about my past challenges, he'd stop me and say something like, "Diamonds can only be made with extreme pressure," or "Stress, pain, and illness are all teachers." He would tell me that my whole life has prepared me to be a catalyst for change. He said my challenging childhood, the pain of lost loved ones, my love of music, mysteries, technology, and science fiction, along with the deep desire to help people learn to thrive, were all a part of a universal plan. This was all a part of my training. He would mention that nothing was a coincidence and that I had something "BIG" to do. Quitting was not an option.

Just before he left the planet, he told me he was handing me a baton. I told him I could never carry his giant baton because my specialty was not civil or human rights. Mr. Gregory was displeased with my response and went totally off on me. He was known to do this when someone asked or said something without thinking. He grabbed my head and said, "You are going to do what I tell you to do. Right now, most people are no longer human or civil. They are physically, mentally, and spiritually ill, completely controlled by parasitic beings. They need a "Doctah B" before they can get any rights."

Dick Gregory is not the only person who has told me that my life and all its challenges were preparation to help people help themselves from a disease of the mind that is consuming the world. The book **Virus Of The Mind,** an essay by Richard Dawkins, is an excellent book on this subject.

Mr. Gregory told me that I would have to teach, create music, write books, make movies, sharing all the gifts I possessed, or my mind would eat me alive. This reminded me of an old statement people would ask when you seemed to be going through something or had a nasty disposition. Hence, the title of this book: **WHAT'S EATING YOU**? He committed to writing the book forward, but he transitioned on August 19, 2017, years before the book's completion.

I miss him, but he is never far away. Mr. Gregory visits me from time to time as an owl. He will always be one of my teachers, my other big brother, my friend, and a member of my Master Mind group.

This book is based on Doctah B's and others' research, personal perspectives, points of view, experiences, stories, scientific studies, and published reports.

Chapter 1:
The Early Days

Growing up in the northeastern USA, I was fortunate to have great parents and wise grandparents who did their best to provide a great foundation. Love, security, guidance, and education were family staples. My parents did their very best, and as far as I know, we never went without food, clothing, shelter, or television.

For me, we embodied "The Good Life". I loved to eat and lived in joyous anticipation of the next meal or snack. Coming from one of the largest families in New England ensured lots of family celebrations, all of which included loads of food, music, and unrehearsed live entertainment.

One of my favorite pastimes was watching Sci-Fi while eating Neapolitan ice cream, colorful sugary cereal, cheesy popcorn, potato chips with some dip, or some other deliciously unnatural, non-digestible snack. Back then, we had no idea how unhealthy our lifestyle was, but at the time, it was normal to me. The TV commercials seemed to jump off the screen, and these short stories informed me of the many products our family needed to make our home a happy, whole, and "normal" part of society. I remember one day thinking that most of the people I watched on TV who had nice things and were living "the American dream" didn't look like me or anyone I knew. Well, except Aunt Jemima and Uncle Ben. In my creative young mind, I imagined they were my distant, wealthy relatives I hadn't met yet. So, I ate as much of their products as possible to support them.

Early on, I recognized that the 5, 6, 10, and 11 o'clock news seemed to be of utmost importance to adults in my life. This thing they called "news" had them "hanging" on to every word these news broadcasters spoke, so I renamed it the noose. The news was the topic of most discussions between my family and their friends. When the winter trench coat-wearing hot summer-wearing newsman, Walter Cronkite, was on the TV, the adults in the room received him as if he were a rock star. Somehow, he magically held their undivided attention.

I wondered how this man could stand the 115-degree heat in Vietnam yet wear this winter trench coat. He was buttoned up tight to the neck and did not break a sweat! I remember bringing this up to my father, asking, "How is this man not breaking a sweat?" I was told to stop being foolish and was sent to my room to do homework immediately. "Focus on getting good grades," he demanded, "grow up to be intelligent." Back then, I thought he was chastising me for joking while the 6 o'clock news was on. I was joking a lot, but this was a serious question. Later, I realized that my mother and father were very perceptive and were trying to teach me valuable life lessons. They taught me there was a time to speak up and a time to be silent, and timing was everything.

Little did I know that our minds, in many ways, were being conditioned to join those who were under the spell of what I now call 'The Daze Of The Weak.' To be under that spell, one must be blindly obedient, submissive, god-fearing, addiction-prone, and a mindless consumer, programmed to accept whatever the network programmer desires. As time went on, things began to feel like we were living in a sci-fi movie like the ones I used to binge-watch.

I loved the shows where alien beings would land in the woods at night and slowly take over the minds, bodies, and souls of defenseless earthlings. These stories ended with the hero falling in love with a beautiful woman and then, together, identifying the alien's weakness, thereby saving the world. I identified with this scenario at a deep level and aspired to be the one who helped save the world.

My family seemed to be trying to live "The All-American Lifestyle," which I called TAAL. We kept a routine Black Middle-American schedule, including Sunday worship services, which I saw as a time to beg God for forgiveness, blessings, and healing for the sick. After church, we gathered to eat and pray, asking God to bless the food. The meals usually consisted of the Standard American Diet (S.A.D.), which was sad, so we prayed over it.

We had no idea our diet was high in crossbred mutant animal fat back then. Our foods were laced with rock salt, highly processed refined sugar, and genetically altered wheat, soy, and corn. We ate overcooked vegetables, artificial sweeteners, poisonous, rancid (rotten), man-made (manipulated) plant oils, and toxic chemical dyes. We were eating things that were eating us alive. Beyond being detrimental to good health, these items were full of flavor enhancers, parasites, piss, microplastics, and preservatives. Even worse, these food pretenders contained nerve-destroying excitotoxins, genetically modified organisms, and natural flavorings derived from roadkill. We literally were consuming products fortified with artificial nutrients and waste byproducts from various toxic industries. Was this why there were so many sick people to pray for at church? Unfortunately, this is not science fiction but a scientific fact.

Most North Americans consume excessive amounts of uric acid (urine) from meat, lactic acid from milk, carbonic acid from carbonated *soft* drinks, oxalic acid from nightshade vegetables, and benzoic acid as a preservative. Excessive acid levels are corrosive to the body and can burn tissues. Instead of creating heaven on earth, people are unknowingly creating hellfire inside their bodies. Most of our meals are drugs dressed up like food being served at a party for the unwitting.

During my twenties, I became successful in the music industry. I lived on the road and in recording studios most of the time. My limited nutritional choices were fast junk foods. On holidays, I'd eat what I thought was a good meal: domesticated white meat protein, white potatoes, white rice, white bread, and some overcooked vegetables. Television had programmed us to think that anything white was right.

A good book on this subject is *DRUGS MASQUERADING AS FOOD by Suzar.*

In the early 1990s, I lived in Los Angeles, California, and I became a successful music producer. At some point, I became depressed even though I had achieved what I thought was substantial financial success. I received many awards, gold and platinum records, Grammy nominations, and Golden Reel awards for engineering excellence. Some would say that I should have been happy, but a harsh reality was eating at my soul. The entertainment industry was more about parasitic relationships, fake transactional friends, perpetual greed, dirty politics, social engineering, unethical commerce, unhealthy competition, modern-day slavery, and metaphysical mind control, and less about true art, creativity, or the creation of harmony on the planet. Whoever controls the broadcast platforms can manipulate the mainstream population by creating harmony, order, chaos, and disorder. Images, tones, rhythms, melodies, and words profoundly affect our mindset. Have you ever wondered who chooses the artists, songs, fashions, actors, and colors programmed for the mainstream public? Do you think the programmers have your best interest at heart, or do you think that people are choosing what's best for them?

I woke up to the fact that I had become a pawn in a massive chess game. This knowledge was eating at me from the inside out. For a moment, I forgot what my mother and father had taught me about timing. As an experienced drummer, I knew that timing was essential; it was indeed everything. But I became angry and started speaking out about the raw realities of the entertainment industry a bit too much, too soon, and too loudly. I found myself in very hot water with entities I call "The Soul Catcher Parasites." I found out the hard way that silence, in some instances, is truly golden because shit got deep fast for me. The fine details of my experiences while in the entertainment industry, I will in no way speak on it here, but I will say that, in my experience, the mainstream entertainment industry is viciously carnivorous, having the power to eat people alive.

My health declined drastically as a direct result of the career I chose. My lifestyle, lack of sleep, food choices, and the mental and spiritual stress I was dealing with took a toll on me. Modern-day doctors claimed they could not help me. My body and mind were at war with each other. This was an autoimmune response. Like many people in North America, I had become allergic to myself, consumed by worms, waste, and worry, a fatal combination.

Day by day, I could feel the life force draining from my body. I felt as if aliens were taking me; I was leaking energy, and like a ship at sea, I was sinking fast. My bank account was healthy, but my body and mind were not. I discovered creative ways to self-medicate. Getting up enough energy to drive to the liquor store or find a street pharmacist to relieve the constant physical pain and deep sadness became my routine.

Chapter 2:
Messenger

One rainy day, I drove down Crenshaw Boulevard in Los Angeles, California, to The Liquor Bank to get some "medicine." I had done this many times before but realized something felt very different this time. I heard the radio announcer say it was March 21st, the first day of spring, the equinox. It was strangely different; there were hardly any cars on the road, which was unusual during rush hour in Los Angeles.

I recall an eerie vibe in the air. Things seemed to move in slow motion, like a dream. The liquor store was empty except for the man behind the counter, who usually greeted me warmly. I spoke to him, but this time he never looked away from the TV. He stared with a blank expression. I thought I was reacting to all the self-medicating I'd been doing recently. I wondered if I was getting high just by thinking about getting high.

As I reached for a bottle of liquid relief, I could feel someone staring at me. All I wanted was my liquor, but I was moved to see where the energy was coming from. As I turned and looked, to my surprise, I saw this beautiful, dark chocolate-skinned woman standing in the aisle watching me. My eyes met hers, and, without any greeting, she asked, "What do you think you're doing?". She had this deep, magnetic look in her large, cat-like eyes. I instantly felt that she could see into my soul. I was compelled to answer her, but I had no idea what to say. A cat woman had caught my tongue.

Everything about her was extremely beautiful; her body emitted a glow. She carried a mild scent of Sandalwood, Gardenia, and Vanilla. I've always loved the smell of Gardenia flowers in the rain. She wore an off-white sweater and a short black dress and had the most beautiful legs I had ever seen. She was adorned with jewelry made of crystals and silver. My mind suddenly left my illness and pain. Suddenly, I felt much better. This mystical woman's energy was medicine to me. I felt like I was under a healing spell.

She told me she had something to share, gave me a phone number, kissed me on the forehead, and left the store. I stood in shock for a minute, staring at the phone number. She wrote the numbers in a geometrical hieroglyphic style, like something I'd seen in a science fiction movie. I stood there for a minute, stuck in a daze, wondering if I was imagining all this. Then suddenly, I ran out of the store to find her. She had vanished along with the rain just that quickly. But the magical vision of her was still in my head. Her mystical smell was still in the air. I felt energetic and alive. I hadn't felt that way in a long time, and maybe never.

I forgot all about what I went to the store for and rushed home. I leaped up the steps as if I had never been ill. Without the slightest hesitation, I dialed the artistically written, glyph-like phone number. To my surprise, she answered, "What took you so long to call?" We talked most of the night about my outlook on the world and my recent health challenges. The next day, I called and asked if she wanted to come by my place. She agreed, which began the fastest cleanup job ever, since my home looked like a hurricane aftermath. I could feel the butterflies in my soul fluttering as my heart pounded the rhythm of pure excitement. It's funny that once we have found something new that energizes us, our focus on old ways of thinking or feeling seems to fade. I was feeling good.

She arrived just as the sun was setting. The sky's orange glow reflected through my picture window onto the platinum and gold records on the wall, setting a golden mood. As I opened the door, she walked right past me without saying a word. She stopped in the archway that led to my Moorish-style home's large sunken living room and began scanning the room.

The living room and master bedroom had been converted into a recording studio, complete with drums and instruments from around the world. Amazed, she stepped down into the room and slowly walked around, touching each instrument. She seemed to merge with them. It was like she was from another planet, experiencing something new, like a child. Honestly, I was very nervous and began feeling strange about the silence. After all, I didn't even know her name. She must have realized this and broke the silence by commenting on how clean the place was.

We sat on the large green sofa in the middle of the room, and she began asking me many questions about my life. One after another, she dug deep into me like a detective. Almost magically, I answered each question without even thinking. I felt like I had been hypnotized, placed under a spell. It was not like me to open up like that. At that time in life, I was usually the one asking all the questions. I was excited by her presence, but I also felt strange about this scene.

After what seemed like hours of answering, I asked her, "So what about you? What's your name?" She closed her eyes tightly and said, "This is not about me, and it's all about you." She paused for a long while and slowly began to tell me about herself, and that she was on special medical leave from the military. I asked what happened, and she began to cry. She must have cried for about 15 minutes before saying, "You might think I'm crazy like most people do, but..." Doing my best to reassure her, I said, "Well, they have called me crazy most of my life; they named me Crazy Man in school." She laughed a little and asked if I believed in life on other planets. In my mind, I screamed, "*Yes! I knew she was an alien*"! She did not seem surprised at all when I told her that I had seen UFOs and strange lights in the sky several times while growing up. The lady continued her story, saying, "Well, like my mother, I was chosen to work on a secret UFO project in the military. The others in my unit and I had a close encounter with alien life and some advanced technologies. I emotionally lost it. After this mission, the military doctors attempted to deprogram our minds. But that didn't work for me, so they used electric shock and mind-altering drugs on me." She explained that she was diagnosed with amnesia, fits of rage, and schizophrenia. She told me that she has missing time events, post-traumatic stress syndrome, and was currently on several mind-altering medications to suppress these conditions. Her voice lowered a bit as she said, "When I get nervous or upset, the symptoms return." She also shared with me that she gained intuitive abilities and developed precognition following her encounters, meaning she saw things before they happened.

Suddenly, it was like the music stopped in my mind, and the uneasy feeling I had turned into worry, then grew into fear. I thought *Oh shit, I've got to get her out of here. What if she has an outbreak, a relapse, or a PTSD event and snaps while she's here?* I had seen this kind of thing in the movies, where the beautiful woman turns out to be a monster or just crazy as hell. Then I figured she was probably reading my mind, and I was getting myself into some big trouble. I began to feel ill again and a bit dizzy. What had I been thinking, inviting a strange woman into my home? Even though she was beautiful and seemed to possess magical qualities and high intelligence. I then realized I didn't even know her name.

Before my mind could wander any further, the woman told me that she knew I was sick, and she insisted that the only thing that could save me was natural plant medicine. She extended her hand, offered me two tickets to a lecture on natural health, and said, "Your life is about to change big time. Always remember to follow the path of least resistance, like all of nature does." The statement, especially, moved me and would prove particularly important in my healing journey. I had heard this statement from my father, my favorite Uncle Bee, in electronics school, and a metaphysics book.

She prepared to leave and told me I might not see her again. She gave me the most incredibly long hug, then said, "Learn to love yourself. You are what you eat and think". She walked out the door, and again, I stood in a daze before rushing out the door to find her. Again, she had vanished. I sat on my couch in a cold sweat, reviewing what had happened. Playing it over and over in my head, I wondered whether I had lost my mind or had been hallucinating. I recall thinking that I had died. No one ever answered her phone again. But that couldn't be the case, because I had her card and the event tickets as proof that she was here.

I attended the lecture, which did indeed change my life. After hearing that information, I had no problem letting go of my consumption of domesticated animals, and I began to live a more wholistic lifestyle. I walked away from toxic friends and associates, and I also left behind the stress of the music industry lifestyle. I released many unhealthy habits and beliefs almost overnight! I became totally immersed in everything surrounding plant-based medicines, natural living, wholistic health, and indigenous health practices. I then began to study every type of natural health modality I could find.

I realized that many diseases were caused by the consumption of abused, domesticated, hormone-injected, antibiotic-fed animal products, in addition to eating overly processed foods and fast foods riddled with unnatural chemicals. All these things work against our body's need for homeostasis or balance, causing mental illness, dis-ease, nutritional imbalances, cellular toxicity, and premature death.

Within one year, I went from feeling my life force being drained away and having doctors basically give up on me to being healthier and having more energy than most of the people around me. The best part was that I was doing it all naturally. Most of my time was spent studying; all I could think about was living a natural lifestyle. Whether that meant attending classes, lectures, workshops, or watching hours of videos about natural health, I was all in. Sometimes I would fall asleep at night with books, videos, and research papers scattered across the bed. I would wake up, drink herbal tea, take my natural supplements, and get right back at it. I was driven. That was a trait I learned from my mother, who would passionately research a subject until she found the answers, no matter what.

My new passion was spreading the word about natural health. I realized one day that I had gotten out of hand and became a natural health zealot. Everywhere I went, all I talked about was how bad the food was, how many chemicals were in the food, and what was killing people. Those around me became upset; even though I intended to help, I was aggressive, egocentric, judgmental, angry, and insensitive. Some friends and family members distanced themselves from me, and rightfully so. I had taken it way too far. So, I learned to calm down, speak only to people who asked me about this subject, and let people live their desired lives. Years later, I became a highly regarded natural health educator, plant-based medicine formulator, and researcher in the Los Angeles area.

Chapter 3:
Mystic Voyage

While on a cruise to the Bahamas in September 1996, I had yet another mystical experience that would again alter my life and my perspective on our world. The seven-day trip profoundly opened a whole new world for me in the health field.

My family and I went on a cruise to the Bahamas to celebrate my parents' 50th wedding anniversary. We bought our tickets for a low price. We soon discovered why the tickets were so cheap; it was hurricane season. As fate would have it, we were forced to leave Nassau early due to an incoming hurricane. The cruise liner had to veer south to avoid the storm, which took us directly into the Bermuda Triangle. As this happened, I felt a strong surge of energy while sitting on the bow of the ship, watching the storm roll in on the horizon. I started hearing what sounded like large bells or gongs in the distance. The hair on my arms stood straight up, and I started to feel strange, almost like I had taken magic mushrooms, but I assure you I hadn't. Suddenly, around 1:24 am, I completely blacked out, experiencing missing time. I know it was 1:24 because my Citizen watch is still fixed on that time. I remember regaining consciousness at sunrise, when the sun and moon rose together; both seemed extremely large. Rainbow-colored clouds swirled across the sky, and the ocean was calm and smooth as glass, with no ripples. Feeling totally disoriented, I became overwhelmed with emotion and started crying and screaming. Then I heard a voice in my head say, "Look into the water." To my surprise, thousands of dolphins circled the entire Carnival cruise ship clockwise, and the sight was astounding. Seeing this calmed me down.

I still have no memory of what happened during the blackout. Afterward, I began seeing symbols, shapes, and colors, and hearing sounds in my mind. I thought I had lost my mind once again.

After the cruise, I found myself seeing and writing a sequence of numbers. I was hearing a strange musical melody. And I was experiencing what I thought were both nightmares and daymares.

There were strange aromas that didn't come from my physical surroundings, like the scent of herbs cooking. Although I'm highly skilled at using my nose to identify things, I had never smelled these aromas before.

I was experiencing a paranormal phenomenon known as clairolfaction, or clear essence, also referred to as the gift of psychic smell. I was also perceiving numbers, shapes, and colors, known as clairvoyance, and hearing sounds in a way consistent with clairaudience.

Every day, spiritual experiences are often misdiagnosed as psychosis. I now understand that if I had sought medical help, I might have been diagnosed with conditions like phantosmia (hallucination of odors), confabulation (distorted memories that a person believes are true), or even schizophrenia, and most likely prescribed pharmaceutical drugs to numb the symptoms and make it all go away.

It all felt like a never-ending psychedelic trip through another reality. My childhood science fiction stories had come true, but this time it wasn't just entertainment. For me, it was real. I had no appetite or desire to talk. My family might have thought I had "snapped" again, and I might have.

Since the cruise, I had been fasting only on salad. I had spent three days alone at home, experiencing strange phenomena. It eventually occurred to me that breaking the salad fast could help break me out of my mental state. So, I went to The Good Life Health Food Cafe. This is where I was first introduced to vegan soul food. It was where a host of metaphysical teachers, liberation leaders, philosophers, and a large community of truth seekers gathered and ate together. These were people who lived a more wholistic lifestyle. People came from all around to hear lectures at The Good Life because it was the place to be, a thriving oasis of knowledge in the desert of Los Angeles. There, I honed many skills, including public speaking. I had been giving a weekly lecture series there for 2 years and had built a strong following.

When I walked into the building, I noticed that only the owner, Efa Sade, was present, which was beyond odd. This place was usually packed all day, every day. Having been on vacation for two weeks, Efa Sade was excited to see me. She looked at me and said, "Welcome back, Doctah B! Oh my God, what's happening to you? You're glowing!"

In a shaky voice, I told her I wasn't sure what was happening, but that the cruise ship I was on had ended up in the Bermuda Triangle, and something had clearly happened.

Efa was very excited and hurriedly pulled some large old books from under the counter and said, "You've been through a portal and have had a dimensional shift." I told her, "Look, Efa, I'm feeling very weird and just need to eat some food ASAP." She began cooking the food while talking a mile a minute. She was reading aloud from books about the energy of the Bermuda Triangle, ley lines, energy vortices, and many metaphysical and mathematical mysteries. Usually, I would have been highly interested, but I was in a daze. I could only hear her way in the background of my thoughts.

Meanwhile, in my head, there was constant chaos of sounds, numbers, colors, and these very intense smells. The food only took Efa about ten minutes to prepare, but it seemed like it took forever. She placed the food down and said, "Doctah B, you don't comprehend a word I'm saying, do you?" I shook my head no and began to eat like I'd never eaten before. As I was leaving, she told me I had been "taken," and I needed time to find myself again. "I pray you'll be ok in a week or so. Drink lots of water and put your naked feet on the earth daily to get grounded", she said. I could barely hear her speaking in the background.

She asked me whether she should cancel my Tuesday night lecture, which was scheduled for the next day. I said, "I don't know." I was not in a condition to make plans for a "tomorrow" that seemed so far away.

When I got home that night, I slept until 5 pm the next day. I was experiencing lucid dreams along with hot and cold flashes, all while having the strange feeling that something was trying to speak through me.

I suddenly, without thinking, jumped out of bed, took a shower, and hurried down to the Good Life for my seven o'clock lecture. The place was packed. People knew I was on a trip and were always eager to hear about my eventful journeys.

As if under a spell, I rushed to the stage, grabbed the microphone, and vaguely remember saying, "6000 years ago something fell or landed on the planet and since then we have become host to several types of parasitic beings that are taking control of our minds, bodies, and spirits."

I don't remember the rest of what I said. Apparently, I spoke about mental, physical, and spiritual parasitic entities in detail for 3 hours, although I have no recollection of what I said. I have relied on the accounts of others who witnessed this event.

Although I had never studied parasites in depth, I discussed them in detail. According to those in attendance, I delivered most of this information with my eyes closed and in a strange, monotone voice devoid of my usual lively, animated self.

I was told that after 3 hours, I suddenly stopped the lecture, leaving the audience speechless and shocked. I stood there staring into space until an older lady who always sat in the front row said, "Doctah B, now that you've scared the mess out of all of us, what do we do about this hellish problem?"

Supposedly, I said, "I have no idea." The people became agitated and outraged. This type of reaction is common when shocking information is given without offering a solution.

The security escorted me out to my car for my safety, and I sped off, still in a daze. I fell asleep with all my clothes on and woke up about 24 hours later. I was wet from head to toe with a splitting headache and no idea of what had happened.

For about six weeks, I was lost in my inner space, day in and day out. I was experiencing uncontrolled emotions, confusion, and lucid dreams. I felt like I had lost my mind, and I had. What eventually helped me was Kava Kava, Motherwort, and Tulsi tea, which calmed my nerves a bit and reduced the noise in my mind.

At some point, I took my watch to a repairman, who said the parts inside were fused, as if they had been exposed to intense heat. Here is an actual picture of the watch that I still keep to this day.

Missing Time

Chapter 4:
The Prophecy

Even though I wasn't feeling well, I forced myself to go to a health food store to pick up more herbs. While there, I ran into an old associate who informed me that a famous botanist and Dogon Priest, Dr. Anthony Kweku Andoh, was lecturing at Christ Unity Spiritual Center. I heard Dr. Andoh had vast knowledge of African medicinal plants, and I suddenly felt deeply motivated to see him speak. I hadn't been out in the community since my strange talk at The Good Life Café, but I was driven to go. Dr. Andoh was giving a presentation titled *"The Prophecy On The Wall."*

I arrived at the lecture, slipped in the door, and sat in the back, hoping to go unnoticed. His slide presentation consisted of pictures that his wife, Kali Sichen El, had taken of a strange, graffiti-like projection on a high-tide retention wall in San Francisco. While many thought the image was a graffiti-like painting, authorities reported it to be a projection from somewhere way out in the ocean, but there was more to it than that. He and his wife knew that the message on the wall was vital for humanity.

Dr. Andoh was a well-respected botanist who had conducted remarkable research on plants and mystical phenomena. To my surprise, several of the numerical sequences and symbols in his presentation were the same as the ones I had been seeing in my mind. I was overtaken by emotion and broke down crying. His slide presentation presented an alternate history of life on Earth, one that differed significantly from the stories the ruling class has taught us. It revealed that many beings visited Earth in its early days, and that some people, plants, and animals had cosmic origins. He explained that we were all aliens to Earth, and the entire universe teemed with life.

After his presentation, I was attempting to slip out quietly when Dr Andoh called out to me, "Doctah B, don't you leave without talking to me." I was shocked that he even knew me! He pushed through the crowd that was awaiting autographs and embraced me. He introduced his wife, Kali, who seemed shocked to see me, as if she knew me from somewhere. Dr. Andoh said something about my destiny; he gave me his phone number and asked me to call. He made it clear that it was imperative that I call him the next day at precisely noon. I left wondering how he knew me. To my knowledge, I'd never met him or Kali before.

The next day, I dialed him. Before I could say anything, Dr. Andoh began cursing me out for about 20 minutes with his deep Ghanaian accent. I couldn't understand everything he was saying, but I could feel where he was coming from. He was speaking harshly and cursing at me, and I wondered why. After all, this is our first real conversation since I met him just the day before. Then he suddenly says, "You have no idea what I'm saying, do you?" I declined, and then he explained in a calmer tone that I had been dealing with some deceptive individuals. They offered me a large sum to be the opening act on a national lecture tour. Although it paid exceptionally well, the people behind the tour had a hidden, reptilian-type agenda. They only wanted to utilize my motivational talents and energy to attract a broader black audience. The main attraction on this tour was a prominent author, whom I won't mention here, whose message at the time was very controversial and fear-driven.

Until this offer, I only received donations for my lectures; the most I ever received was $150. They were offering $5,000 per event, plus travel, hotel, and food, and 25 events were booked. This was a big-time opportunity.

My message has always been about helping people take control of their lives, delivered in an entertaining way. I was dedicated to teaching people to become more aware by sharing information that could help them achieve mental, physical, and emotional harmony. The money had blinded me. I was about to allow my life force energy to be used by human parasites to validate a message that was in total opposition to mine, for the money. I had told no one about this tour proposal, and to this day, I have no idea how Dr. Andoh knew about it.

Dr. Andoh said it was time to wake up and learn more about the different types of beings visiting the Earth. Some were friendly, and some were parasitic, only here to use our life force purely for their survival.

When I asked him how he knew me, he explained (if memory serves me correctly) that he and others were having a recurring dream in which I was the focal point. He went on to say that I was here on the planet to help people eradicate physical, mental, and spiritual parasitic beings. I asked if he meant worms in the digestive tract. But he meant systemic parasites that lived throughout the universe, in the bodies of nature, and other parasites that had infected the minds and spirits of humans and animals. Shocked, I asked if he meant aliens. He replied, "Some are extra-terrestrial, some are terrestrial, some are inter-terrestrial, some are energetic and spiritual." According to Dr. Andoh, even generational beings use the life force energy of entire families, communities, and nations.

This reminded me of something that the Indigenous people often spoke of, which has many names, including Wetiko or Watiko, Windigo, Wintiko, Wendigo, Antimimon Pneuma, or "counterfeit spirit." It is the flyer and Vampire Virus of the Mind. It short-circuits people's connection to self, the tribe, or the group and promotes colonies to collapse. It causes people to self-sabotage, live in fear and anger, and self-destruct. It is a cancer of the spirit that eats its host alive.

Watiko is a contagious psycho spiritual parasite of the mind.

Symptoms:
Greed
Jealousy
Envy
Chronic Selfishness
Chronic fear / Anger

I told Dr. Andoh that this subject was deep and far beyond my knowledge, experience, and expertise. But he insisted, saying, "You have what it takes, and you've been training all your life for this. During your recent trip on the ship, you were downloaded with instructions that will unfold in time." I asked him how he knew all of that, and he calmly said, "The first thing you need to learn is that the knowledge about everything is in every molecule there is. You will learn how to tap into what's already there, hidden from the eyes of the blind. You hear and feel a vibration and a rhythm that is unique to you, "Doctah B

Sirius." I said, Doctah B, Sirius? He replied Yes, Doctah B Sirius.

That was the first time I heard the name Doctah B with "Sirius" attached, and I liked it! He told me to study the Sirius star system and that when I noticed it in the night sky, it would blink red, white, and blue while sending waves of information. After he told me this, I found myself staring at Sirius for hours at a time. I soon became very connected to Sirius, the brightest star in the sky.

"Have you figured out the formula yet?" Dr. Andoh continued. "What formula?" I asked. He replied in a strong Ghanaian tone, "Don't bullshit with me, young man; we know you've been working on a new formula for parasite eradication. I'll be in town next week; bring it to me."

Over the years, I have met many people with special powers and abilities. At times, I feel like I am on the Starship Enterprise, the ship used by the Federation of Planets, "to seek out new worlds and go where no one has gone before" from the 1960s Star Trek TV series. I constantly meet people with unique talents, explore new ideas, seek more profound knowledge, see strange things, and go where others could not or would not go.

I have witnessed things that most people consider impossible or even supernatural. This was one of those moments. Dr. Andoh was correct. Since my channeled presentation at the Good Life, I'd stopped lecturing and started working on a natural protocol for systemic parasites. I was being intuitively guided by what felt like cosmic forces as I created this unconventional blend of herbs, essential oils, and trace minerals.

When Dr. Andoh returned to town, we met in the parking lot of an L.A. health food store called Simply Wholesome. I handed him the herbal formulation in a brown paper bag, and he held it without opening it. Then, he closed his eyes and said, "You're almost there. Three herbs are missing." I said "What three herbs? He quickly replied, "It's your destiny, not mine; only you will know the answer."

He said, " I can tell you this much. "One is from Egypt, one is a native American herb, and the other is from the Caribbean. Find them, and you will have the formula that will help people eradicate systemic parasites. You will discover how to assist people in releasing the mental and spiritual parasitic entities." He went on to say, "You are here to help certain people awaken from a deep 6000-year sleep." I thought this thing about sleep was a bit funny because I would love to sleep for 6000 minutes (4 days). It had been challenging for me to sleep lately. After all, I still felt the effects of the Bermuda Triangle experience.

At the time, I was having a profound awakening; my senses were heightened, and the sounds, colors, tastes, and feelings were intense. I'm sure if I had told others about what I was going through, they would have wanted me to see a psych doctor who would most likely have me committed and or drugged to "shrink" my brain and turn it all off. In some cases, that might help, but I was being guided by Dr. Andou and a feeling in my gut.

I was feeling driven like never before. I began researching everything I could find, and within three months, the three herbs I had previously been unaware of appeared in unexpected ways and seemed to be what was needed for the formulation to be complete.

One day, I was cleaning up my room when my arm hit the bookshelf. A book fell to the floor and opened to a page about a Caribbean man who used wood from a tree high in quinine to treat people with illnesses such as malaria, caused by parasites in the genus Plasmodium. I realized that this must be the Caribbean plant medicine needed for the formula.

Soon after, I was at a health food store and saw Chief Little Bear, an indigenous medicine man who had become one of my teachers. We spoke briefly, and then he reached into his pocket, pulled out a root, and asked me to take a bite. It had a strong smell and a pungent, weird flavor. It quickly opened my sinuses and lungs, and I felt lightheaded. He explained that this powerful root is called Bear Root because bears eat it as soon as they emerge from hibernation to kill the worms that have grown inside them during the long winter's sleep. He said that when consumed, it causes the bear to emit a vapor that confuses wolves' smell receptors. The bears are thin and weak when they first emerge from hibernation, and they remain defenseless and vulnerable for weeks. For this reason, they choose areas to hibernate where this root is plentiful, protecting them from attack when they awaken. This root enhances the bears' vision and senses after they have been in darkness for so long.

Indigenous people and other plant teachers also use this root during a vision quest for spiritual guidance and insights. It's also used for parasites, spiritual protection, eradicating demons, and boosting the immune system. I knew it was the indigenous American herb I needed for the formula.

Not long after, I received a phone call from a man who had heard me on an earlier radio interview on Stevie Wonder's radio station KJLH in Los Angeles. Carl Nelson and Jamaal

Goree hosted a talk show called The Front Page, on which I had been a regular guest for many years. This show is how I became known to the public in 1994. The man on the phone said his name was Muhammad and that he had information he felt would be very useful to me. However, he didn't want to discuss it over the phone, so we agreed to meet.

When we met, he said very little. His skin was dark with a hint of royal blue, and his eyes were very dark. He possessed a strong mystical energy. He stared deep into my eyes while he handed me a handful of little black double-sided pyramid-shaped seeds. He said they grew near the Nile in Egypt and had been mentioned in the Quran, stating they cured everything but death, and he quickly walked away.

For months, the seeds sat on my counter, and I had no idea what they were until I walked into my favorite Indian market, New India Sweets and Spices, on Fairfax Blvd in L.A., and there they were, in a bag on the spice rack, for sale. I grabbed the bag, went directly to the counter, and asked my friend what they were. He said something in an Indian dialect that I couldn't understand. On a piece of paper, he wrote the words kala jeera and said, "Get no sick." I realized then that this was the third plant medicine needed to complete this unique formula.

From then on, I would invest many hours studying and experimenting with ways to combine those herbs with the ones I already had to create a program that would help eradicate systemic parasites safely.

When a person focuses intently on an idea or outcome, the Reticular Activating System (RAS), an essential part of the brain, automatically searches for evidence to support that idea, belief, or outcome. This is where **the Law of Attraction** concept comes from. We are naturally magnetic and attractive, drawing to ourselves more of the thoughts, words, and actions we invest time and energy in. The RAS system is like a genie or faithful servant following the orders of the captain of the vessel: your mind, body, and spirit. This system's sole purpose is to seek out what we command, regardless of the circumstances. Yes, our daily thoughts, words, and actions are commands to our entire being-ness. We are all scientists, researching our environment to prove our mindset. The RAS system is connected to every part of the brain and nervous system; it is essential for life.

They pop up everywhere when we consciously focus on agreeable outcomes and opportunities. Daily thoughts of Love, Harmony, Good health, Wealth, Inner Peace, Joy, and Abundance create powerful attractor waves that vibrate into the quantum field and eventually materialize into reality.

Conversely, when our daily focus is on disagreeable outcomes, obstacles, and problems, the mind searches for evidence to support the idea of more of them. Not realizing this leads many of us to wonder why we constantly have bad luck, misfortune, hardships, and setbacks, and why we feel cursed; we may just be attracting it. Have you heard people say things like:

"I can't win for losing." We are predicting loss, so it appears.

"Stupid me." This statement predicts and attracts stupidity.

"They make me sick." We are predicting illness.

I'm in no way saying that if we think positively, nothing bad will happen. I'm saying that temporary challenges and setbacks will happen in life. How we respond and adapt to life changes will determine whether we overcome or become victims of our thought patterns, which consume us like mental parasites.

Our brains can predict, so it may be best to be optimistic and imagine the best-case scenario rather than the worst-case scenario. Keep in mind that our thoughts and words can create a self-fulfilling prophecy. The saying "Energy flows where the mind goes" is very accurate!

I was so inspired by Dr. Andou that my thoughts, words, actions, and daily goals became focused on creating an adaptogenic herbal program. I imagined and predicted that it would help eradicate systemic parasites safely while balancing the body's major systems. I learned that many products used to eliminate intestinal parasites are harsh and tend to deplete the body's vital forces, making the parasites and superbugs even stronger. Wise formulators know that certain herbs can be harmful if combined or misused; they must be mixed and curated just right so that the outcome is favorable and beneficial for healing the whole being.

This was uncharted territory; I had to listen carefully to my intuition and improvise, just as I do when creating new music. Frequency and vibration were coming into play as I realized that herbs were encoded with specific colors, textures, and smells. However, these things also indicate their healing properties, and they began to reveal specific sounds, even entire songs, that I could hear. Then I had a major revelation: individual plant medicines have voices and make sounds like musical instruments. Similarly, in music creation, I applied the same concepts used in studio mixing by amplifying specific plant energies and reducing others to create balance, while ensuring the main melody was clearly heard. The melody in the herbal formulation must become the focus or goal of the formula, just as in a good song.

When using plant medicines, it may be wise to remember that much of the mainstream reference material on herbal medicines is limited to the beliefs, perceptions, culture, politics, and goals of the people gathering the information. I have noticed that much of the research documented by Westerners about Eastern herbal medicines has little to no concept of the countless ways that the indigenous people used medicinal plants, especially on a spiritual and emotional level. They are often only looking for the plant's basic use and/or what they perceive as its active ingredients. They usually have no idea of the plant's metaphysical properties or of the soil and original environment in which it grew. This can make their conclusions flawed or at least incomplete.

Here are a few reference guides that document the spiritual and metaphysical properties of herbal medicines:

The Science And Romance Of Selected Herbs Used In Medicine And Religious Ceremony By Dr. Anthony K. Andoh

Herbal Secrets Of The Rainforest By Leslie Taylor

Herbs And Things By Jeanne Rose

Cunningham's Encyclopedia Of Magical Herbs By Scott Cunningham

Using my Intuitive gut feelings versus documented information became my guide, as I could now hear the musical language of certain plants. I was not aware of any reference materials that even remotely suggest that plants emit sound.

The book *The Secret Life of Plants by Peter Tompkins and Christopher Bird,* and the documentary of the same name, are the closest. They demonstrate that plants respond to sounds, thoughts, and words, and they highlight how plants generate electrical impulses in response to their environment. I was tapping into something profound and being led back into music, sound, and vibration. Plant sounds and vibrations may have been utilized by ancient civilizations but never documented because some of this information was passed down through oral traditions and culture.

I realized more each day how right Dr. Andoh was in knowing that I had been prepared for this mission for most of my life, starting with my career as a musician. I realized that everything in the universe is based on vibration. We are tiny universes, and the vitality and frequency of every cell must be in a state of equilibrium; otherwise, we will become out of tune, also known as illness. The proper use of medicinal plants, nutrition, and mental and physical exercise can move us from imbalance to balance and create order out of chaos. I was very focused on my goal, and I attracted everything I needed to develop a program to help people reach the goal of parasite eradication while restoring balance to their lives.

Dr. Andoh soon returned to Los Angeles, and we met once again. He grabbed the bag that included the three new herbs, closed his eyes, never opened the bag, and said with a huge smile, "This is it." Then he hugged me. I still wonder how he could hold the bag and know if the herbal combination was correct.

He then emphasized that these creatures, known as parasites, are ancient and exceptionally resilient. They have excellent survival skills. "You must be mindful to change the combination and ratios of certain elements every time the seasons change, or some parasite's offspring will become immune to the product, and others will mutate, becoming even more difficult to eradicate. This is why most products used against parasites lose their effectiveness over time. It's up to you now, Doctah B Sirius."

Over the years, I gained massive respect for Dr. Andoh. He helped me realize my new purpose and redefine myself!

I was experiencing a significant turning point in my life. Until then, I felt I was a successful musician, music producer, herbalist, and public speaker, but now I was experiencing a quantum leap. I had suddenly shifted from a crowded path in the world of health and wellness to a whole new frontier one that was mine alone.

After conducting extensive research, I discovered that parasites are among the leading causes of disease, along with Candida albicans overgrowth, dehydration, nutritional deficiencies, stress, and chemical toxicity. When I met Dr. Andoh, I had been an herbalist for about 6 years, gathering knowledge of traditional herbal medicines used to help people become healthier. My research up to that point had focused solely on eliminating common parasites in the digestive system. I now realize that the most detrimental type of physical parasite is the systemic kind, which lives throughout the body. I also began studying mental parasites that obstruct the thought process and spiritual parasites that suppress the spirit.

It dawned on me that most of the people on the planet were being eaten alive by physical, mental, and spiritual parasites. My new mission was to help people seeking to become or stay healthy realize what was eating them and teach only those willing to break free.

I have faced many challenges, but my parents' 50th-anniversary Carnival Cruise trip has marked a new life path for me. I'd gone through a complete transformation, where I was forced to lose my old mind and go through Hell to create Heaven on Earth. My life purpose was now more explicit than it had ever been. I realized that some of the fictional hero characters that entertained me in my youth were merging with who I am now, one who helps people recognize and eradicate what's eating them.

At the time, I was experiencing several of the steps of the Hero's Journey described by Joseph Campbell in his book The *Hero With a Thousand Faces*.

The ordinary world - My life growing up, TV, and snacks.

The call to adventure - Experience on the ship and the aftermath.

The refusal of the call - Extreme self-medicating to escape what I had experienced on the ship.

Meeting the mentor- Meeting Dick Gregory and Dr Andou in the same year.

Crossing the threshold - Realizing my new path.

Return with the elixir creating the parasite protocol.

Chapter 5:
What Is A Parasite?

The information I am about to share is of the utmost importance. Our very survival depends on knowing what could be eating us alive.

We must apply specific protocols to eradicate and neutralize parasites while restoring homeostasis in our lives in the most natural way possible.

A traditional parasite is an organism that relies on another organism for survival. It lives in or on another organism, known as the host. The host provides energy, nutrition, protection, transportation, and a place to reproduce. Parasites that live inside another creature are called endoparasites, and those that live outside or on the surface of the host are called ectoparasites.

The etymology of the word "parasite" comes from the ancient Greek and Latin word *parasitos*, "one who eats at the table of another." *Para* 'alongside' + *sitos* 'food'.

These beings are uninvited guests, like someone who might move in with you and pay no rent, buy no food, and bring no supplies; they live off you for free. Parasites constantly consume, absorbing all their nutrients and energy from the host and giving nothing in return. They can drain life force energy, absorb essential nutrients, and manipulate the host's activities.

In humans, physical parasites range from microscopic to 35 feet long in the intestinal system. Parasites, or worms as they are sometimes called, are among the most abundant life forms on Earth, and many types live in colonies and groups. They are also among the oldest beings in the animal kingdom, intelligent masters of survival. Some parasites reproduce at alarming rates, laying more than 5,000 eggs per day. Other types reproduce without even laying eggs; some divide and replicate. When cut into pieces, several varieties of parasites make a new worm from each severed piece. Some bacterial and viral types, by contrast, replicate.

Some would argue that viruses and bacteria are not classified as parasites, but my research goes beyond these traditional assumptions.

Parasites are a natural part of life on Earth. They inhabit a wide range of environments and live in and on plants, animals, insects, water, soil, water vapor, and even dust particles. They are excellent at adapting and transmuting and proficient at survival. Some create protective dwellings and enclosures known as cysts to survive harsh environments. Certain types can enter suspended animation (stasis) to survive extreme temperatures above 700 degrees Fahrenheit, and some are even found alive in ancient ice samples. They are crucial for the planet's survival. In some cases, parasites facilitate the decomposition of dead animals and plants, returning them to the soil as fertilizer. Other parasites help clean up debris.

Parasites should not be confused with friendly bacteria or intestinal flora that assist in the digestion of food and make up a large portion of the immune system. A healthy gut microbiome helps protect the body from dangerous levels of disease-causing bacteria, harmful viruses, toxins, and certain types of parasites. Maintaining healthy intestinal flora is essential for good digestion, clear thinking, a balanced immune system, and overall well-being. The average person today has an imbalanced gut microbiome, which is incredibly unhealthy both physically and mentally.

When the host population of parasites is low, there may be no noticeable change in the host's health or behavior. However, as the population grows, the host may experience physical and/or mental challenges. At this time, illness, disease, and imbalance will likely appear in the host.

Parasites are one of the leading causes of illness in the world, along with yeast (candidiasis), dehydration, nutritional deficiencies, toxins, and stress. Parasites absorb the nutrients from the foods we ingest; they deposit their waste into our bodies, invading our sacred space. This process can cause organs to fail. This invasion alters the host's internal chemistry and disrupts its natural functions. Many parasites live, eat, and mate within a 28-day cycle, following the lunar cycle's circadian rhythm. Chronic infestation weakens the host's natural defenses by inhibiting hormone function and proper elimination, interrupting digestion, and triggering inflammation. Malnutrition may arise at this point, causing issues in the body that were previously thought to affect only people from distant lands, but not in North America.

In school, we were taught that parasite problems mainly existed in third-world countries. This was a common belief and a dangerous assumption. These disease-causing pests live everywhere on Earth. The overabundance of parasites in humans is a modern-day plague, a true pandemic. It is estimated that these bugs and their toxins have infected up to 80% of the population. Parasitic worms are some of the most toxic agents known; they are among the primary underlying causes of disease today. In medical practice, parasitic infections are often misdiagnosed or left undiagnosed, leading to inappropriate treatment. Illness caused by medical examination, treatment, or mistreatment is called iatrogenesis and is said to be the 3rd cause of death in North America. This information is published in The American Medical Journal.

Parasites, nutritional deficiencies, oxidative stress, preservatives, highly processed foods, chemical toxicity, widespread insecticide use, and current factory farming practices make our bodies toxic waste dumps. The human population has become the perfect breeding ground for life-threatening parasites, viruses, and fungi.

Parasites also prevent the body from healing from diseases, which may not have caused the infestation, resulting in countless physical and psychological health challenges. Parasitic worms may be the underlying cause of the spread of Hepatitis, Cancer, Diabetes, Obesity, Heart Disease, sexual dysfunction, Addictions, Cold / Flu, and mental imbalances, among many others.

The mainstream approach to these illnesses has been to treat the symptoms using pharmaceutical drugs or surgical procedures, which often have many side effects and may not deal with the root causes. Research has shown that this approach may weaken the body and exacerbate the parasite problem, ultimately leading to the development of superbugs over time. Superbugs are created when organisms mutate and become resistant to chemicals. This is a common problem in agriculture and the raising of domesticated livestock. More potent versions of these chemical pesticides and fungicides are used each year to control insects and diseases that can destroy food crops.

At the same time, these organisms, in an attempt to survive, produce offspring that become increasingly difficult to eliminate as they develop immunity to the chemicals. These poisons end up in the air, are absorbed into the soil and the water table, creating a toxic waste dump that pollutes the water we drink, cook with, and bathe in every day. To add to all of this, genetically modified organisms or GMOs, growth hormones, and antibiotics are commonplace and acceptable in food production. Naturopathic research has shown that inorganic chemicals may not be the most effective approach to addressing this epidemic. Could this be the reason that, over the last 50 years, our immune systems have become significantly compromised? Have we created a parasite paradise?

Many parasites are zoonotic, meaning they can be transmitted easily from animals to humans and from humans to humans. Humans have a weird relationship with animals. They love their pets; they kiss them, eat with them, live with them, and even sleep with the animals they call pets. Most pet lovers are unaware that these animals can cause illness. They have become a part of the family and are often treated better than human relatives. There are pet hospitals, hotels, boot camps, and graveyards, and most pet foods have more life-enhancing nutrients than what's on the human's dinner plate.

At the same time, modern humans torture, kill, and eat more animal flesh than ever before. The slaughterhouses, feedlots, and factory farms are sheltered behind high walls with razor wire and armed guards. I wonder why? Most people would like to blame corporations when they find out how their food is being produced. But the public drives the demand for more meat, so it's simply a case of supply and demand.

According to an NPR Morning Edition article from 2012, the average American consumes 270.7 pounds of meat annually. Research indicates that there are approximately four cows per human globally. Is the rainforest being cut down to make room for more cattle? If there are at least 7 billion people on the Earth, then do a little math and think about where all this domestic animal waste is going. Keep in mind that domesticated animals are not naturally occurring; they are produced through animal husbandry, selective breeding, and cross-breeding. The waste from these mutants and clones is not naturally recycled back to Earth like that of natural animals; it's highly toxic and poisons the planet at an alarming rate.

The documentary Cowspiracy states that modern-day factory farming is the number one cause of pollution, not automobiles and other industries. Imagine billions of animals overcrowded, sick, suffering, and living in their waste while passing bacteria, mold, viruses, parasites, and diseases back and forth to each other and into the environment. Would you knowingly eat zoonotic worms for dinner? Billions of people do just that by consuming domesticated meat, sometimes three times a day. These people are slowly being eaten alive from the inside out. Current research shows this may be the source of many plagues, outbreaks, and pandemics.

It's a good thing that millions of people are waking up and choosing to be more conscious consumers. Increasingly aware people are choosing natural, non-domesticated animals or adopting a vegan or vegetarian lifestyle. If this continues, we may have enough time to save some of humanity.

The average person may consume infested, chemically laden, crossbred, preservative-rich, moldy, processed, and denatured foods every day of the year. This weakens our immune system, allowing parasites to thrive. When these toxins are eliminated from the diet and the body is cleansed, healing begins by strengthening the body's natural defenses.

Remember that parasites and microbes are not just found in commercially processed foods, but also everywhere, including in the environment. Vegetarians and vegans carry them too. Cross-contamination and poor hygiene practices also facilitate the spread of parasites and diseases from one organism to another. This can occur through contact with animals, bodily fluids, sweat, and even casual contact. One trip to a public restroom, a restaurant kitchen, or a surprise visit to a good friend or relative's home could give you a new perspective on cleanliness. Some people rarely wash their hands; some don't bathe very well or often enough, if at all. Others may use a "wet rag" under their armpits and then bathe in cologne. People cough, sneeze, scratch, and pick their noses and other body parts, only to greet you with open hands that have not seen water or soap. When you push an elevator button, grab a handrail or a doorknob, and even handle money, you are exposed to parasites and a host of microorganisms. Don't take it personally when some folks don't shake hands. I've learned to bow a lot. This may sound funny, but it's the truth.

The Battle for Domination

Old-school biologists thought simple parasites, bacteria, and pathogens could never intelligently manipulate our lives so profoundly. This flawed theory has been another significant misjudgment of Western society. However, many indigenous people were aware of these creatures' threats and used herbal formulations, ceremonies, and traditional practices to keep the threat level low. Of course, many scientists who spent years using corporate-backed research dollars in the past would never agree that parasites could control human activities. That line of logic turns out to be an egotistical mistake that costs the lives of millions of people each year. The truth is that what they considered primitive and uncivilized people knew more about life on Earth than most modern scientists do.

There is an emerging branch of science called psycho-parasitology or neuro-parasitology. It is the study of how parasites influence the thoughts and actions of their host. Yes, Mind-controlling parasites. An excellent book on this subject is *"This Is Your Brain On Parasites" by Kathleen McAuliffe.*

No matter what man believes, parasites may rule our world. They control the way we live, even manipulating societies. They have much to do with how we think, feel, mate, eat, and sleep. These parasitic beings shape our belief systems and traits.

There has always been fierce competition between hosts and parasites a never-ending battle for dominance.

Humans have a long history of stories about pathogens decimating populations. Over time, people have adopted new ways of living to protect themselves. Both parasites and hosts evolve and develop new ways to survive. Humans have acids on the skin, in tears, in the mucus of the mouth, inside the nose, lungs, and digestive tract as the first line of defense against disease. Our immune system is always searching for anything that could make us ill.

Parasites have several advantages over humans. One is that they are approx. 2.5 billion years old by some estimations and have developed ways to survive that man has yet to comprehend. Two, they are among the most plentiful beings on earth. They can also replicate, modulate, mutate, and adapt to thrive under harsh conditions. As you can imagine, this gives them the upper hand in terms of survival.

As I mentioned earlier, certain parasites can alter the behavior of their hosts. The host ultimately helps transmit parasites from one to another. They eat foods that the parasite wants them to consume. Many foods weaken the host's ability to defend itself against parasites. The host's natural instincts are disrupted, impairing their judgment, gut feelings, or intuition. This leads the host to lose its natural reasoning, unknowingly helping the parasite achieve its goal of survival at any cost.

Certain parasites, such as bacteria and viruses, are the result of human and scientific advancements. There have been numerous cases of infectious agents and pathogens "leaking" from laboratories into the general population, causing outbreaks. Some superbugs were designed to be bioweapons in war, and others were intended as ethno-specific pathogens made to destroy only certain groups of people deemed "undesirable," a type of ethnic cleansing. Anthrax is one example of such a bioweapon. This scenario is discussed in the Military Review, November 1970, in the article *"Ethnic Weapons: Race-Specific Biological Weapons."*

This type of "Franken-science" has caused mutations and created active micro-monsters in our environment, which have been passed from animal to animal, person to person, and from person to animal, polluting our waterways and lands. Many disease epidemics are a result of man's foolish quest for power, population control, domination, xenophobia, and pure greed.

Our worst threat may not be the nightmarish fears of snake bites, shark attacks, spider bites, unsettled ghosts, terrorists, or man-eating monsters. Those are nothing compared to the parasites in, around, and upon us. These are the real aliens; some come from outer space. Why else would they put astronauts in quarantine upon returning from space? Why are we told not to go near areas where space rocks, satellites, or meteors fall? Some scientists theorized that the first life forms on Earth came from meteors and visitors from other worlds. We are not alone and have never been!

Let's Meet the Parasites

Parasites come in many forms microscopic, visible, physical, mental, and even energetic. They can live in food, water, soil, air, and sometimes inside our very thoughts.
They enter through what we eat, drink, breathe, and believe.

Some are ancient biological hitchhikers that have evolved alongside humanity, learning to hide, adapt, and feed off our nutrients, minerals, and mental energy. Others exist in the unseen realms, energetic or emotional parasites that thrive on fear, distraction, and disconnection from spirit.

While some infections are silent, others reveal themselves through fatigue, anxiety, cravings, inflammation, confusion, or loss of inspiration. Each parasite has its own agenda, and each teaches us what happens when we lose sovereignty over our internal terrain.

The Tapeworm Family: The Ribbon Eaters

Tapeworms are among the best-known intestinal invaders. Their flat, ribbon-like bodies can grow up to **33 feet** inside the intestines, absorbing nutrients directly from the food you eat.

This is why many people infected with tapeworms remain hungry no matter how much they eat because the parasite gets fed first.

Tapeworm larvae can migrate beyond the gut through the bloodstream to the **liver, lungs, and even the brain**, sometimes forming cysts mistaken for tumors.

As they mature, tapeworms release egg-filled segments (proglottids) that exit with feces, contaminating soil and water. Birds, fish, and livestock consume these eggs, continuing the cycle of infection that moves up the food chain into humans.

- **Taenia saginata** – found in beef

- **Taenia solium** – found in pork

- **Diphyllobothrium latum** – found in fish

Common Tapeworm Symptoms:
Nausea • Weakness • Diarrhea • Abdominal pain • Constant hunger or no appetite • Weight loss • Fatigue • Vitamin and mineral deficiencies • Seeing segments or thread-like worms in stool.

Taenia solium:

May cause **Neurocysticercosis**, leading to headaches, memory loss, confusion, seizures, and, in extreme cases, death.

Giardia lamblia: The Invisible Water Invader
Thrives in untreated water; survives regular chlorination. Causes severe diarrhea, cramps, and nutrient loss often misdiagnosed as "stomach flu.

Cryptosporidium: The Waterborne Trickster

Originates from cattle-farm runoff. Resists conventional water treatment and weakens the immune system, causing chronic intestinal distress.

The Fluke Worm: The Stealth Assassin
Flatworms are found in pork, chicken, fish, and beef. They invade the liver, pancreas, lungs, and bloodstream, disrupting organ function.
Linked to cancer, diabetes, and autoimmune disorders.

Entamoeba histolytica: The Flesh-Eating Amoeba
Causes **Amoebiasis**.
Symptoms: Bloody diarrhea, abdominal pain, fever, nausea, weight loss, cancer and sometimes abscesses in the liver or brain. Many carriers show no outward signs.

Loa Loa: The Wanderer
Spread by deerfly or mango-fly bites through the skin and eyes, creating a visible crawling sensation.
Symptoms: Itching, swelling, eye pain, fatigue, and cognitive changes.

Rapid die-off can trigger brain inflammation (encephalopathy).

Pinworms: The Night Time Crawlers
Tiny white worms that lay eggs around the anus while you sleep.
Symptoms: Anal itching, rash, restless sleep, irritability.
Eggs spread easily via bedding and surfaces, surviving up to three weeks.

Whipworm: The Colon Feeder
Burrows into colon tissue and feeds on blood.
Transmitted through contaminated soil or crops fertilized with human waste. It can lead to chronic diarrhea, anemia, and even colon cancer.

Strongyloides: The Muscle Madness
Lives in muscle tissue and spreads through skin contact with contaminated soil.
Symptoms: Stomach pain, alternating constipation and diarrhea, rashes, and muscle weakness.

Leishmania: Desert Syndrome
Carried by sandflies.
Causes **Leishmaniasis**, known for skin ulcers and organ damage.
Linked to **Gulf War Syndrome**, manifesting as fatigue, mood imbalance, and muscle pain.

Tuberculosis: The Death Cough
A bacterial parasite that hides inside white blood cells.
Symptoms: Persistent cough, fever, night sweats, weight

loss, and exhaustion.
Spreads easily through coughing or talking.

E. coli 0157:H7: The Hidden Food Parasite

A deadly strain found in contaminated meat, milk, and produce.

Symptoms: Severe cramps, bloody diarrhea, fever, kidney failure, and death. Wash your hands and food carefully; contamination spreads fast.

Pfiesteria piscicida: Cell From Hell

A shapeshifting marine parasite linked to toxic algae blooms and ocean "dead zones."
Releases neurotoxins, causing memory loss, respiratory distress, and neurological damage.
Tied to runoff from industrial pig and poultry farms.

Listeria monocytogenes: The Lunch Meat Marauder

Found in soft cheeses, deli meats, and poultry.
Survives freezing and heat.
Especially dangerous for pregnant women and infants.
Symptoms: Fever, nausea, confusion, miscarriage, and in severe cases, death.

Trichinella spiralis: Porky The Worms

Causes **Trichinosis**.
Symptoms: Muscle pain, facial swelling, fever, diarrhea, death, and fatigue.
Larvae migrate to muscles and can persist for years.

Schistosoma mansoni: The Blood Fluke
A hermaphroditic parasite is transmitted through contact with contaminated water and snails.
Lives in blood vessels and damages the liver, kidneys, and intestines.

Blastocystis hominis: The Modern Mystery Parasite
Once thought to be limited to developing nations, it is now widespread.
Symptoms: Gas, bloating, rashes, IBS-like pain, fatigue, and immune weakness.

The Multi-Layered Symptoms of Parasitic Infection
(Physical • Mental / Emotional • Spiritual / Energetic)
Parasites feed on more than food they feed on energy, attention, and vitality. They can drain minerals from the blood, dull mental clarity, and distort emotional balance.
Recognizing the full spectrum of their influence is the first step toward liberation.

Physical Symptoms: The Body as the First Battlefield

Digestive & Nutritional:
Constant hunger or no appetite • Weight changes • Bloating • Gas • Cramps • Constipation/diarrhea • Anal itching • Teeth grinding • Drooling in sleep • Sugar cravings • Food sensitivities • Bad breath • Coated tongue

Muscular & Skeletal:
Aches • Weakness • Joint pain • Stiffness • Twitching • "Crawling" sensations • Chronic soreness

Immune & Inflammatory:
Frequent colds • Allergies • Rashes • Eczema • Swollen lymph nodes • Low-grade fever • Autoimmune flares

Neurological & Sensory:
Headaches • Brain fog • Dizziness • Light sensitivity • Tingling • Restless legs • Poor coordination

Endocrine & Reproductive:
Hormone imbalance • PMS • Infertility • Low libido • Sexual dysfunction • Prostate or vaginal infections

Circulatory & Organ Stress:
Anemia • Puffy eyes • Irregular heartbeat • Blood-sugar swings • Liver tenderness

Sleep & Fatigue:
Waking 2–3 a.m. • Night sweats • Chronic exhaustion • Low morning energy

Skin & External:
Itchy scalp or ears • Rashes • Acne • Brittle hair/nails • Premature aging • Body odor

Mental / Emotional Symptoms The Mind as the Mirror of the Gut

Cognitive:
Brain fog • Forgetfulness • Poor focus • Confusion • Mental fatigue

Mood & Personality:
Irritability • Depression • Anxiety • Mood swings • Hopelessness • Emotional hypersensitivity

Behavioral Patterns:
Addictions • Compulsive habits • Procrastination • Victim mindset • Self-sabotage

Social / Relational:
Isolation • Distrust • Codependency • Energetic drain after socializing

Nervous-System Stress:
Overreacting to small triggers • Feeling "wired but tired" • Random fear or dread • Low motivation

Spiritual / Energetic Symptoms The Soul Under Siege

Energetic Signs:
Heaviness or fog • Loss of inspiration • Inner criticism • Feeling drained after certain people or places • Resistance to healing • Guilt or shame loops • Chaotic dreams • Nightmares • Sleep paralysis •
Cold or tingling sensations • Dimming aura • Attraction to toxic people • Repeating karmic loops • Feeling "fed upon" by negativity

Higher-Dimensional Effects:
Blocked intuition • Creative stagnation • Psychic interference • Difficulty manifesting goals • Attraction to parasitic systems (energetic, social, spiritual, personality defects)

The Unifying Thread

The body, mind, and spirit form one sacred ecosystem. When any part becomes toxic or unbalanced, parasitic energy moves in physically, mentally, or spiritually.

Parasites don't only infect; they reflect mirroring the places in us that need healing, self-respect, and light. When awareness shines in, darkness dissolves.

Through detoxification, right thought, aligned emotion, clean food, and elevated frequency, parasites lose their power. Healing is not just removal it is reclamation of your life force.

Your thoughts, words, and actions are the ultimate warriors against parasites.

Revelation Major

There is one more parasite that I would like to discuss that I have a personal connection to: the dog and cat heartworm, Cuterebriasis (Dirofilaria immitis). This parasite lives in the hearts of "man's best friends," dogs and cats. It can grow to be a foot long, and when in a group or colony, it looks like a clump of spaghetti noodles.

When a Heartworm larva reaches a human lung, even if it never matures, it can cause heart and lung disease. These young heartworms die in the lungs, causing inflammation and scar tissue as the body attempts to eliminate the heartworm larvae. This inflammation leads to chronic illness and several life-threatening conditions. Frequently, doctors are unaware that the root of these illnesses is heartworms. The diseases caused by heartworms are called heartworm-associated respiratory diseases **(HARD)** and cause long-lasting damage to the heart and lung tissue.

Some researchers believe this parasite is transferred to humans when a mosquito bites a dog and then bites a human, injecting the parasite. My question is: if the parasites live in heart and lung tissue, why is it concluded that the tiny larva is only transmitted through the blood, not through the pet's saliva or breath? I guess it's better to blame the little, lowly mosquito than the beloved dog or cat. This blame game has happened many times throughout history, and the wrong culprit has been condemned. Sound familiar?

The medicine veterinarians use to treat heartworms is Ivermectin.

Millions of people suffer from illnesses originating from the pets we love, sometimes even more than we love ourselves. Was this information taught in elementary school? Would we still live so closely with animals?

Symptoms

- inflammation, a condition known as pulmonary dirofilariasis
- fluid around the lungs (pleural effusion)
- round coin-like lesions, sometimes called Granuloma. They show up on chest X-rays that could become cancerous.
- Chronic bronchitis
- abnormal cough,
- coughing up blood,
- pain in the chest,
- wheezing,
- chills,
- fever.

- buildup of fluid around your lungs (pleural effusion)
- round lesions that show up on chest X-rays

This puts a whole new spin on the old saying, "sick as a dog."

As I delved deeper into this parasite, I had a significant personal revelation. Growing up, my family had a Boxer named Major; he was a very big dog. My brothers and I all slept in the same bedroom, where Major loved to visit whenever he could. He usually lived outside or in the basement when it was very cold outside until our parents were away, and then we did as we pleased. As long as I can remember, Major drooled and had a cough like a soft bark, unlike his usual ferociously loud bark. Major was our protector; everyone in our neighborhood feared our big dog, Major.

A peach tree was growing near the back porch beside Major's doghouse. One day, my oldest brother, Greg, grafted a plum branch onto a peach tree as a science project while I was watching. As I write this chapter, it hits me: My oldest brother Greg had inadvertently given me my first lesson on cross-breeding, also known as hybridizing, by splicing that peach tree on that spring day in March. I'd check the spliced branches my brother wrapped in twine almost daily. By mid-August, we were beginning to see the fruits of his labor. It grew a weird fruit, and my mother strongly urged me not to eat it. This is the kind of stuff that still excites me to this day. I became a researcher when I was five or six. All this is just coming to me now as I write these words.

Many years later, as an educator, I occasionally highlight the advantages and disadvantages of plant crossbreeding in workshops and lectures. Hybridization and crossbreeding produce a new organism from two or more original species. They are widely used in agriculture to control pests, domesticate (denature) crops, and increase yields (profits).

The peach/plum tree has become another important topic for me as an herbalist. Peaches, plums, cherries, and apricots are all very nutritious fruits, but their pits or seeds have impressive health benefits. They contain high levels of Vitamin B17 (Amygdalin). Vitamin B17 has been used as a natural medicine for thousands of years. It has been used successfully in both cancer prevention and treatment. It helps to boost the immune system, balance blood pressure, prevent oxidative stress, suppress parasite growth, and control inflammation, pain, and free radicals.

Laetrile is a synthetic drug made from purified B17. It was made illegal in the 1980s, citing that it was poisonous and that there was no proof it helped as a cancer treatment. After much success with natural medicine, they made all forms of vitamin B17 and B15, as well as unprocessed apricot pits illegal. After many years of scrutiny, the organic versions of the apricot pits, known as bitter almonds, are now appearing in health food stores.

Little did I, as a boy, know that the peach tree from my childhood possessed a powerful medicine that I would one day incorporate into an alternative cancer treatment. I now call this peace tree The Peace Tree.

Okay, back to Major. Suddenly, I remember lying on the ground next to Major under the peach-plum tree and the evening sky, staring at the stars. I especially remember this bright, twinkling star; I now know it to be Sirius, the brightest star in the sky.

When Major made his transition, the veterinarian said he died from a heart/lung disease and old age. They said he had chronic bronchitis, which affects the smaller airways in the lungs. It dawned on me that my brothers and I have had challenges with chronic bronchitis for most of our lives. Despite undergoing numerous herbal, mineral, and enzyme treatments, cleanses, and even fasting, I still experience a slight cough and occasional mucus in my chest. The doctors said it was scar tissue from childhood bronchitis. Now it all makes sense: this issue is the long-term side effect of heartworm-associated respiratory disease (HARD) caused by the heartworm of the only dog I ever loved Major.

Then I began thinking of what Dr. Andou said about the name Doctah B Sirius. I remember him, my mother, my father, Chief Little Bear, and Dick Gregory all saying that there were no co-incidences and that my life's purpose would become more evident as time passed.

Sirius is the star I'm named after and is also called Alpha Canis Majoris or "A Dog Big." It's the brightest star in the sky in the constellation Canis Major. Canis Major means Big Dog. For me, all this is a major revelation!

Chapter 6:
It's All In Your Head

In her book **"This is Your Brain on Parasites," Kathleen McAuliffe** addresses a new branch of science known as neuro-parasitology. This is the scientific study of parasites that manipulate humans and animals by influencing and ultimately taking over their neurological systems, natural functions, and host actions. It is now known that these superbugs have played a significant role in shaping human societies. They affect belief systems, education, finances, politics, and habits; they encourage xenophobia, division, race relations, and wars. Old-school scientists did not want to accept that parasites could control their hosts' thoughts and habits. The arrogance of these old "know-it-all" scientists would not even consider that these little bugs have so much intelligence and influence on human life. Many traditionally trained scientists, even when presented with new facts, will never admit they could ever be wrong.

The parasites themselves seem to be controlling the minds of the scientists, politicians, influencers, and those who fund them, the so-called 1%.

Brainworms are parasites living in brain tissue. They manipulate brain function and the host's natural thought process. A parasite is an organism that relies on other organisms for necessities and survival. Brainworms are highly evolved biological manipulators, aliens to the host, that disrupt the host's behavior so the parasite can achieve its goal of survival, multiply, and thrive.

Remember, parasites use the host for food, shelter, and transportation. Transportation includes traveling from one habitat to another or from one stage of life to another. The parasite's survival depends on how effectively it alters its host's behavior and immune system. It spends all its time achieving its goals: staying undetected while keeping the host disorganized, disoriented, and somewhat mindless. When the parasite is in control, the host constantly makes choices that only serve the parasite.

Rabies is a parasite that infects its host through a bite. Many people have contracted it through an infected dog or other animal bite. It takes over the mind of the host and can cause delirium, hallucinations, abnormal behavior, anger, fear, paralysis, and foaming at the mouth. This parasite is also known to make its victims very aggressive and violently mad. Sometimes, the host will scream wildly, yell, growl, and even howl like a wolf. It makes the host want to attack, wound, and bite other warm-blooded beings to travel from creature to creature. The symptoms can appear anywhere from a few days to up to a year after contact. The average is 1 to 12 weeks, depending on how close the bite or wound is to the brain. Also, the host size may determine how quickly symptoms appear. Up to twenty-six thousand humans contract Rabies each year. This is the origin of the Vampire stories.

A fly called **Apocephalus Borealis** injects eggs into the abdomen of bees. After being infected, these bees act strangely, then abandon the hive and become what's termed Zombie Bees. This can cause Bee Colony Collapse Disorder (CCD), a phenomenon where an entire population abandons its home. It isn't only Apocephalus Borealis that threatens bee survival; it also includes people parasites. This parasite is destroying Earth's ecosystems by using pesticides, fungicides, and toxic petrochemicals on a large scale. These two parasites are responsible for the drastic reduction in bee populations. Note that without bees pollinating plants, the entire Earth will become a desert.

Hookworms can ingest up to 1 cc of blood from the intestine each day, which may explain the widespread anemia. Anemia is a leading cause of many health challenges, and blood quality dictates quality of life. There is an American variety of hookworm called "Necator Americanus," which means "American Murderer." Why? Because it kills Americans! More than 644 million people worldwide may be infected with hookworms.

In the early 1900s, an epidemic of hookworms in the American South caused widespread anemia. Symptoms of anemia include weakness, headaches, irritability, difficulty concentrating, mood swings, sluggishness, dizziness, lightheadedness, tingling in the extremities, cold hands and feet, and slurred speech.

Around this same period, malaria reached plague levels in the South until the swamps were drained. Some scientists say that may be how slow speech and a southern drawl developed, amid the scourge of hookworm and malaria. It is said that this combination affected the brain's frontal lobe, where speech is produced, causing the slower, slurred speech of people in the South. Affected parents spoke this way, and their children learned to talk by modeling their parents and passing it down. It is also believed that microbes worldwide play a significant role in how different cultures develop, whether through microbe manipulation or humans adapting over hundreds of thousands of years to survive parasite infections.

Malaria is caused by a parasite called Plasmodium. When mosquitoes bite infected animals, Plasmodium is transmitted to the mosquito. Although it is only a single-celled organism, it is a powerful manipulator of its host. Once inside the mosquito, the parasite reproduces in the gut, and after approximately 7 days, its progeny reach the infectious stage. They then manipulate the mosquito's natural behavior, causing it to crave more blood than usual. These infected, super-hungry mosquitoes now bite as many animals and humans as possible to complete the parasite's mission: mass infection and reproduction. It can also change the odor of infected humans, attracting more mosquito bites while the person is ill. This helps it hitch a ride to the next host quickly. In humans, it promotes blood flow by modulating platelet function. This allows more blood to be drawn during a bite. This hitchhiking parasite is responsible for more deaths than any microbe except its relative, Toxoplasma.

Toxoplasma Gondii - The Mind Manipulator

This single-cell parasite is the most notorious of all, sometimes referred to as the queen or king of parasites, and soon you will understand why.

Toxoplasma begins its life cycle in the gut of domesticated cats. The cat may not show any symptoms of an infection, but the parasite's cyst ends up in the cat's waste. The feces and urine of infected cats are irresistible to rodents, dogs, and many animals; they even play in and ingest it. These animals are naturally very cautious, but this parasite makes them lose their instincts. Remember the old saying, "When the cats are away, the mice will play." Play in what? Cat poop and pee!

Once ingested, Toxoplasma travels to the Amygdala, a region of the brain. The amygdala is a pair of almond-shaped organs within the brain. Their essential function is to collect signals from all the senses, process differences between healthy and life-threatening stimuli, and simultaneously activate the appropriate responses to promote optimal functioning. Concepts like love and fear, peace and anger, good and bad, friend and enemy, and joy and despair must be sorted out correctly, or confusion ensues, resulting in conflicted thinking.

Malaria was once thought to be the number one killer of humans. Still, Toxoplasma may have taken over the top spot by causing people to self-harm and ultimately self-destruct individually and as a collective.

This parasite can now be found in waste, soil, dust, gardens, unwashed fruit and vegetables, and contaminated drinking water. How often do we test tap water? Most people on the planet have a weakened immune system due to living in a fear-based society, depression, the overuse of antibiotics, a toxic environment, and consuming the Standard American Diet, or S.A.D. This is happening 1440 minutes a day, 365 days a year, keeping people in perpetual protection mode, a mental war zone. When a person is in chronic protection mode and survives, their blood flow is diverted from the forebrain (used for logical or analytical thinking). The digestive system and healing slow almost to a halt. The body's resources (typically used to boost immunity) are directed to the rear brain and muscles, producing high levels of adrenaline, cortisol, and acid when a person is ready to fight or run away from something threatening.

Now, to add to all this, the average person is consuming animal products that were raised in stressful environments, basically tortured, imprisoned, and pumped full of chemicals each day. You are what you eat!

Toxoplasma is a significant cause of The Pandemic of Self-Harm and Mass confusion, confusing the host's emotions while destroying its natural instincts to thrive. This bug can also urge people to engage in harmful activities, unhealthy relationships, and consume toxic foods, committing what I call Nutritional Suicide. Toxoplasma disrupts entire cultures by causing its host to deviate from the healthy traditions that keep them strong. When enough individuals become infected, they may drift away from the safety of the group, tribe, village, or family. Could this be a factor in the shift away from thriving matriarchal societies in which the Divine Feminine was held in high regard, as in bee and ant colonies? The goals of successful cultures are grounded in homeostasis and in what's best for the group as a whole. History shows us that when societies become too male-dominant, they fall out of balance, which is the basis for all diseases and disorders.

It is easy to control the mindset of divided beings or cultures living out of order in a state of mass chronic imbalance. This can lead individuals within the group to become overly self-serving and egotistical, thereby creating conflict. If unchecked, the group will reach a tipping point, ultimately causing the entire colony to collapse.

Toxo can manipulate emotions and cause counterproductive competition, abuse, violence, greed, narcissism, fear, anger, confusion, bipolar disorder, separation, jealousy, materialism, egoism, divisiveness, and toxic vanity. It dumbs down its host to increase its chances of survival. It can cause a type of Mass Schizophrenia. This is a severe mental illness that interferes with people's ability to think clearly, manage emotions, make decisions, and relate naturally. Sounds like a lot of people today. What do you think?

Toxoplasma can cause impaired cognition, slow reaction time, chronic depression, emotional distress, disillusionment, despair, and confusion. It can also cause its hosts to self-sabotage goals they naturally desire to reach. This parasite manipulates the host's innate aversion to predators and other potential threats. Typically, most successful beings possess an innate response to life-threatening things. They naturally move away, take flight, run, protect themselves, and shift into survival mode when a threat is present. Instead, the Toxo-infected brain automatically selects the opposite of what is best for the survival of the individual and its kind, thereby limiting the ability to thrive. It also impairs vagus nerve function, thereby reducing the host's ability to follow gut feelings. It can urge people to create scenarios that keep them from evolving into higher-consciousness beings.

Wherever you find mice and rats, you see cats. I was told by someone who worked at a feedlot that many rats and cats are found in commercial livestock factory farms. He said that the floors of some of these places are often covered with urine, vomit, maggots, roaches, rats, mice, and cat droppings that the livestock ingest. Each time a person consumes domesticated animal flesh, are they risking exposure to Toxoplasma? What about the countless other disease-causing parasites, bacteria, viruses, growth hormones, antibiotics, and toxic chemicals that may be used in factory farming practices? This is an excellent reason to seek out high-quality organic food, which is becoming easier to find nowadays. I have heard several times that this superbug helps to keep the livestock domesticated. Once Toxoplasma establishes itself in the host's brain, it mentally domesticates the host, so it never attempts to escape captivity. This may be a significant factor in why consumers are sick and unable to organize in ways that could change our condition on Earth. The host loses its natural purpose: to thrive and live freely. Are you aware that domesticated cats are not natural beings? Mysteriously, domestic cats appeared in ancient Egypt (KMT) about 4000 to 6000 years ago, theoretically through crossbreeding of wild varieties. It has been surmised that only domesticated cats can carry this superbug larvae. Most domesticated animals have imbalances and carry zoonotic parasites and illnesses not seen in wild varieties. It has been said that KMT held cats in high esteem and had lots of them. Could Toxoplasma be a significant factor in the fall of the original KMT society?

Today, we can see the effects of Toxoplasma in our self-destructive society. Anthropophobia, the fear of people, xenophobia, and the fear and hatred of people perceived as being different are all considered normal in our supposedly intelligent society. Today, there is rampant greed, constant war, obsessive rumination on a disagreeable past, pessimism, speeding texting & sexting while driving, rage, domestic violence, and all of the isms being promoted and broadcast 24/7 on artificially intelligent 5 G and soon 6G networks to devices held in our hands.

Growing up, they called American beer Rat Piss. I asked why and was told that rats often got into the giant vats used to make some American beer and peed in them. I even found some references to this online. If true, this could be another way Toxoplasma could infect chronic beer drinkers.

There is a label warning on kitty litter products. The following is a warning from the product **Fresh Step Crystals Cat Litter.**

Kitty litter Label Warnings: We would like to remind our customers, especially pregnant women and those with suppressed immune systems, that cat feces can sometimes transmit a disease called Toxoplasmosis.

Therefore, always wash your hands after handling each cat litter.

For further information, consult a physician.

Do not flush this cat litter down the toilet.

Do not use this product as a traction aid, as it becomes slippery when wet.

From MSDS: Dust may irritate eyes. Inhalation of dust may irritate the nose and throat.

Prolonged contact of dust with skin may irritate due to a drying effect.

Cat feces can sometimes transmit a disease called toxoplasmosis.

Pregnant women and immunosuppressed persons are most susceptible.

Always remember to wash your hands thoroughly after handling used cat litter.

Used cat litter is not recommended for use in the garden.

The Centers for Disease Control and Prevention *considers toxoplasmosis to be a leading cause of death attributed to foodborne illness in the United States. Severe toxoplasmosis can damage the brain, eyes, and other organs. It can develop from an acute Toxoplasma infection or be reactivated from a prior infection. Severe cases are more likely in individuals with* weak immune systems, though occasionally, even persons with healthy immune systems may experience eye damage from toxoplasmosis.

FDA Website Statement on Toxoplasma:

Did you know that you could feel healthy but still have toxoplasmosis? This foodborne illness is caused by the parasite Toxoplasma gondii.

Toxoplasma gondii could be harmful to you and your baby if you become pregnant.

Here are some frequently asked questions about toxoplasmosis.

"What is Toxoplasma gondii?"

It's a parasite found in raw and undercooked meat, unwashed fruits and vegetables, contaminated water, dust, soil, dirty cat litter boxes, and outdoor places where cat feces can be found. It can cause an illness called toxoplasmosis, which can be particularly harmful to the individual as well as to an unborn fetus. You can become exposed to T. gondii by accidentally ingesting contaminated cat feces, which can occur if you touch your hands to your mouth after gardening, cleaning a litter box, or touching anything that comes in contact with cat feces. Over time, the parasite can enter your bloodstream. It usually takes about a week. If you become pregnant while the parasite is still in your blood, it can pass through the placenta to your unborn child. You can also get toxoplasmosis by eating raw or undercooked meat or drinking water contaminated with T. gondii. So, you should eat thoroughly cooked meat.

Toxoplasmosis can be challenging to detect. However, symptoms typically include swollen glands, fever, headache, muscle pain, or a stiff neck.

Sourced from: https://www.fda.gov/Food/ResourcesForYou/HealthEducators/ucm082328.htm

Chapter 7:
Dreaming Awake

When I was a boy, I had many vivid dreams. From time to time, I would have a recurring vision state in which massive translucent objects descend from the sky, carrying beings that can shift between physical and nonphysical states. In dreams, these beings aim solely to feed on human psychic energy by collecting plasma-like substances from certain people. These visions returned and intensified after my experience in the Bermuda Triangle.

I asked some of my teachers about these dreamlike visions, and they all offered the same explanation. Basically, they said I was channeling past, present, and future events. They all stated that this information would prove valuable for my future mission.

In my dreams, there were many different types of beings living on and within the Earth. Some came from elsewhere in the universe; others shifted through time and dimensions. It is even theorized that we may have come to Earth from elsewhere. When you look at nature, you see millions of different life forms, all living in a state of equilibrium; otherwise, we would cease to exist.

In dreamtime, I was told that between 4 and 6 thousand years ago, objects fell or landed in specific places on Earth, like Egypt and several locations between the 33rd parallel and

19.5 degrees north and south. This is where many pyramids, mounds, and sacred sites are. It is said that nonphysical beings, psychic energy parasites, arrived on the planet; one of their many names is The Soul Catchers. They travel from planet to planet, even jumping across dimensions and timelines to create discord among other beings, then harvesting the resulting disagreeable energy as a food source. They developed the ability to generate Mass Conflict Frequencies that agitate and manipulate minds to consume the inharmonic essences or vapors emitted by people living in chronically stressed states of consciousness. They created a type of human agricultural project, or a worldwide energy farm, in which stress and confusion are the key ingredients that domesticate most of the population all the time.

It has been said that the Toxoplasma parasites may be biological drones that implant themselves in people, keeping the population agitated and under a spell, creating a virtual mental farm. These drones or satellites may be remotely controlled by pirates who now live off the world, traveling in hive-like ships that consume energy from afar. Some may perceive this as evil or say these beings are devils, but it is exactly what humans do. In fact, they do this to the animals and people they deem lower than themselves, raising them as livestock. Using techniques like mass-distraction, deception, distortion, deprivation, cognitive dissonance, crossbreeding, bondage, emotional pain, and chronic stress, people are easily kept under control. Then, man absorbs their life-force energy as food, all in the name of survival. So, are we evil devils, too?

Researchers have theorized that up to 70% of the world's population may be infected with Toxoplasma, one-third of the world's population! What we know for a fact is that Toxoplasma, or whatever controls it, is a factor in the collapse of our human colony.

What's been eating you has been eating all of us for thousands of years. As bad as all this may sound, there are simple solutions that have been used by naturalists all over the world that can inhibit the power of this superbug. I created a specific 28-day herbal program. Pharmaceuticals and Herbs alone are not very effective because they can't pass the blood-brain barrier. Most traditional parasite cleanses have little to no effect on Toxoplasma, Superbugs, brainworms, and organisms engineered as biological weapons.

We use a particular group of Medicinal Master Mushrooms, Essential oils, Flower essences, Crystal elixirs, rare earths, M-State elements, and Minerals from the Great Barrier Reef.

Historically, this type of natural program has been very effective in reestablishing our 28-day circadian biorhythms, which are essential for a properly functioning immune system. When our minds and immune systems are imbalanced or weakened, disharmony or disease results. At that point, we become more susceptible to becoming a host to beings that take advantage of our weaknesses and siphon our energy.

Parasites rely on the same 28-day circadian moon cycle for survival, just like we do. They feed, mate, multiply, and travel at specific times within this natural cycle. The key is to interrupt the archaic superparasites' life cycle and take back your life.

Parasites are considered master manipulators. Their prime directive is to stay cloaked and in control. To continue succeeding, they must influence their hosts to choose foods that weaken them, thoughts that consume them, and willfully engage in daily activities that further weaken their immune systems.

Systemic parasites are brilliant manipulators and masters of survival. They hide in immune-privileged sites, such as the brain, eyes, joints, and other sensitive tissues. They also create slimy coatings called biofilm to encapsulate, cloak, and protect themselves from detection and harm. This makes them totally resistant to most pharmaceuticals and common herbal preparations. Parasites also use this biofilm to attach themselves to tissues, metals, rocks, and plastics. Using microplastics, they form a super biofilm that creates a super-strong protective layer or cyst-like container, also known as a plastisphere. This is one of the main reasons that simple parasite-cleansing herbs are largely ineffective: they cannot degrade these complex structures.

I am sure parasites like Toxoplasma and others have been manipulating humans to engage in self-harm for a long time. They have effectively urged us to poison our own bodies and our environment through our complete dependence on plastics, toxic chemicals, and highly processed foods. This insanity has disrupted our immune and endocrine systems from functioning properly, allowing parasites and microbes to thrive.

All parasitic beings have a soft spot an Achilles' heel. Their survival depends on the host's lifestyle, suppressing their immune response or natural defenses.

Like most life on Earth, they are affected by the moon's circadian cycle and the 28-day human biocycle. It starts with the new moon or low ocean tide. This circadian rhythm is like a drumbeat or a pulse wave, influencing all living things on

Earth. When the body's environment is inhospitable to parasites for at least 28 days, it disrupts the parasite's feeding, mating, and rest cycles. This will cause the parasite to die, leave, or enter quiescence or hibernation. This is true for physical and mental parasites. This is why I created Doctah B's Total Body Para cleanse Program, a 28-day systemic parasite-detoxification and rejuvenation program. I'll share more on this later in this book.

Chapter 8:
The Mycelium Connection

Much research suggests that humans and all life forms arrived on Earth as Mycelium spores or fungi, possibly on asteroids, space dust, or space pods. We share at least 65% of our DNA with mushrooms, which calls into question the weird Darwin theory. By the way, if it weren't for the mycelium and their flowers and fruiting bodies, which we call mushrooms, life as we know it could in no way exist. I believe that the mycelial network, omnipotent and omnipresent, existing timelessly everywhere in the universe, is what humans may call God the Creator of Life. It is in and around us; nothing can live without this superconscious quantum information network or hyper-verse.

My and others' theories suggest that the mycelium network is part of the Quantum Field, the Unified Field, the Akasha, the creator of what we call life.

It is said that we are made in the likeness of God. "Cut open a piece of wood. I Am there". "Break a stone; I am there." From the highest mountain to the deepest sea, I AM there. In every organ of the body, I Am there. Mycelium does indeed walk or float on water. They said God is in the sky; well, mushroom spores are. They told God performs miracles, heals the sick, and raises the dead. Well, mushrooms do all this every day. The Mycelium can and does transform water and some fruit into wine. We were told in church that God was everywhere and there was no place God was not! Mushrooms compost dead matter, transform it, and effect a kind of resurrection, giving it new life as something else, so nothing truly dies. It sounds like the mycelial network is God like to me.

There are more varieties of mycelium than can be counted, and all have specific Pacific roles in all organic life cycles. They do alchemy, changing substances from one state to another, even converting inorganic materials to organic compounds. Some types convert sunlight into energy and vitamin D. The Chaga Mushroom helps feed and repair human melanocytes, or melanin-producing cells. These centers support healthy sleep cycles, the pineal gland, THE FIRST EYE, our cosmic antenna, and a hyper-dimensional vision device. All plants rely on mycelium to break down minerals into small particles that can be easily absorbed. Certain varieties are a source of healthy protein, others are potent healers, and some even have a unique ability to recognize non-human DNA, such as that of Toxoplasma, and eradicate it. They do this in both the plant and animal worlds, acting as powerful detoxifiers and regenerators. We can in no way exist without them!

They act as Earth's internet, facilitating complex communication networks among all organic life forms. They, along with gut bacteria, make up the gut microbiome of all humans and animals. They give us that gut feeling that urges us to choose to become the best version of ourselves. The gut, the Solar plexus, is the most extensive bundle of neurons in our bodies. In my mind, the gut is the first brain, instantaneously receiving subtle information from our local environment, the earth, and the cosmos. This is only achievable when our gut microbiome is healthy, which, in most cases, is not due to our habits, choices, and unnatural lifestyles.

Most people today are out of touch with nature and Mother Earth. Just the act of **earthing, or grounding,** walking barefoot on the earth, helps reset our biorhythms. This discharges chaotic static electrical charges out of the soles/souls of our feet into Earth's massive mycelium network. When this is done just three times a week for 14 minutes, she (Mother Earth) responds by sending harmonically balanced micro-bursts of healing energy throughout our bodies, normalizing multiple organ systems, balancing the solar plexus, and beginning the process of repairing the microbiome and the neurological system. This simple act can help create an optimal state of being that the universe urges us to do.

When we do a Seasonal Systemic Parasite Detox like mine, ground regularly, periodically remove the subconscious trash, replenish our gut microbiome using a 32-strain or more probiotic, and focus on the destination we choose daily, we take complete control as the captain of our vessel. This process has the power to eradicate or neutralize any parasitic beings or pirates by throwing them overboard.

Throughout recorded history, people have used **Sacred or Magic Mushrooms** and other Entheogens as spirit medicine to take what may be called a Vision quest. This would be done to reach other dimensions, heal the mind, or receive answers to more profound questions about life. They are called Master Entheogens for a reason. The word Entheos means "God within." Let that sink in.

Governments made them illegal, stating that they were hallucinogenic and had no medical use. Once again, they lied. The experiences are in no way hallucinogens, which implies they are not real. Instead, they free us from the 3rd-dimensional world and open one up to the quantum universe while performing what I'd call a Psychedelic or spiritual detox. Most cultures on Earth have used entheogens to induce truly spiritual experiences. It is well documented that ancient societies used these plants and other substances to deepen their connection to the spirit realm, traverse dimensions, and experience higher states of consciousness while accessing universal or Divine Knowledge. Some of the people who created what would become religions were channeling sacred information while ingesting entheogens.

Most people who have taken mycelium say they met God," The Most High," inside the mycelium network that exists inside all of us. Psilocybin magic mushrooms and other Plant Entheogens, also known as Plant Teachers, are finally being accepted and used in psychiatry to help treat many psychological issues very successfully.

Whether you call them mycelium, mushrooms, or fungi, they are the galactic transceivers that send and receive all messages instantaneously outside of time. There is no place where the mycelium network is not! All plants and animals depend on them for life; without them, there wouldn't be soil as we know it, drinkable water, or even a livable atmosphere on Earth. Now, this sounds like the PRIME CREATOR!

Side note: Always remember that sacred things should be done at sacred times and in sacred places, or the results could be less than desirable. If you choose to partake in a mushroom vision quest, it should be conducted in a sacred or guided setting to achieve optimal results.

Chapter 9:
A Small Bittersweet Bite Of History And The Western Medicine Machine

Historically, traditional herbology and medicine, and plant-based remedies have been used effectively for prevention and treatment. The fact is that that was all that was available until recently. Herbal medicine helped the world recover from outbreaks and plagues in the past. Herbs and natural substances have been used in foods as flavorings to slow spoilage, prevent illness, control worms, and promote wellness. People all over the world have used plant-based herbal medicine and thrived.

Have you ever wondered why mainstream Western medicine does not support proven and mostly inexpensive natural therapies? It is incredible how many wealthy people who have serious health challenges can afford the so-called best and are still sick after expensive mainstream medical treatments. Why is there no talk about natural health care in the discussions about "health care"?

It seems that today, the term "Health Care" only applies to therapies, practices, services, and pharmaceuticals approved by systems that may not honestly care about our good health.

Western medicine has made significant advances in some areas. Surgery, emergency care, diagnosis, and other forms of treatment have significantly improved. Increasingly, physicians are open to natural alternatives, innovations, less invasive practices, and compassionate care.

In my opinion, the old-school medical industry must step up in the areas of prevention, nutrition, and detoxification. The 100-year-old health model must be upgraded if the goal is to help people truly become and stay healthier. There is mounting evidence that the pharmaceutical machine and marketing firms rule the medical business, not the doctors per se. It looks to me like the mainstream healthcare system is more about profits than healing people.

I had a client who had been diagnosed with a rare cancer. He was prescribed a very tiny little white pill that cost $10,000.00 a month and was told he would have to take it for the rest of his life. He was happy because, with his co-pay, he only had to pay $2,000 out of pocket each month. Wow!

He and his wife had come to speak with me about natural alternatives. I asked for the name of the medicine, but they couldn't even remember it. The husband looked in his bag and realized he'd left it at home, out of town.

They both began to panic. I calmed them down by having them breathe deeply for a bit, and then I asked what the medicine was supposed to do. They had no idea. They did rattle off a long list of side effects that he'd experienced since he'd been taking the $10,000-a-month pill. I gave both of them a shot of my central nervous system tonic called **Heaven On Earth**, and they instantly began to smile and relax.

The gentleman says he hasn't felt this good in years. His wife grabbed his hand and rubbed it gently just as a tear ran down her face. I explained the benefits of plant-based medicines and discussed the symptoms of poor nutrition, including toxins and physical, mental, and energetic parasites. They said they'd never heard this information before. The wife says, "I can't believe our family doctor never told us about these natural alternatives?" I told her not to blame the doctor and that he may not be familiar with this subject, as he most likely never learned it in medical school. I asked whether they had heard of the Flexner Report. They replied no.

In the early 1900s, the Carnegie Foundation commissioned Abraham Flexner to create a report about medical schools and hospitals. This was cloaked as a concern for the quality of healthcare education in the USA. The story goes that the Rockefeller and Cargill drug companies used this report to fund only those in the medical business who exclusively prescribed pharmaceutical drugs. Guess whose drugs?

Historians say this created an all-powerful political and medical coalition that suppressed natural remedies, treatments, and holistic medicines, many of which were highly effective. This created a dictatorship-type medical monopoly that vilified any and all-natural healing methods, whether they were effective or not. The average person was forced to comply, became brainwashed, and became afraid of natural treatments. They surrendered their innate spirituality and lost their discernment and rationality.

They say the Flexner report was created and used to coerce and scare much of the world. They crafted written "proof" that this new monopoly was the best way to ensure quality health care.

People who administered natural medicine and those who used it were often forced to hide out or jailed for practicing medicine without a license. Many natural healers and spiritual guides have been labeled nut cases, snake oil salesmen, witches, sorcerers, and demons. Lots of people have historically gone missing, been tortured, burned alive for exercising freedom, living naturally, and breaking religious and or corporate laws.

According to history, a few greedy parasites virtually eliminated all independent medical schools, drugless doctors, hospitals, clinics, healing houses, and natural health centers. They lobbied, lied, and made-up laws and statistics that made it impossible for those who were not white and male to even go to medical schools. The report went so far as to suggest that black people should only learn to work in hygiene departments because they were unclean and had more germs and parasites than white people. Now that's low down and dirty.

Because of racism and sexism, the Flexner sham blocked up to 50000 or more black potential doctors and women from practicing medicine. When a black health community is being served by mostly doctors and nurses who don't look like them, the community exhibits more stress as a whole because of a history of mistrust and mistreatment. This causes stress on the community as a whole, whether black or white, which decreases the chances of healing and thriving for all people. Black people live longer in areas with more doctors who look like them.

It's very obvious that Flexner's racist report was designed to improve healthcare outcomes for white people only while making drug companies filthy rich. Just imagine the advancements and innovations that could have been achieved worldwide if it had not been for global narcissism and greed.

I could see how some uninformed people in the 1900s might have gone along with this shakedown, but Medical Racism and Sexism are nasty diseases that affect all people everywhere, even to this day.

Today, the doctor you think of as yours may have a strong incentive to take direction from the arm-twisting pharmaceutical marketing firms that suggest what's best for the drug company's profits and not the patients. Does *your* doctor inquire about the whole you, your environment, your state of mind, and your emotions? Did you know that even using the word cure could get you serious jail time even today? Does this sound gangster to you? Well, it gets even deeper.

According to the American Medical Association, the 3rd leading cause of death is MD-directed treatments and what they call medical mistakes. This is called iatrogenesis, defined as adverse effects, illness, or death caused by medical treatment. The United States Department of Health and Human Services reports that 15000 Medicare patients die each month from treatment.

So, is it true that people buy medical insurance based on the fear that if they get sick, they may not be able to afford treatments and be left to die? Does this mean that a large portion of a person's earnings goes to pay dearly to have access to a system that is the 3rd leading cause of death in the USA?

I grew up hearing that I needed to find a good wife, a real, honest, respectable job with a good healthcare plan. This meant I should discover this excellent job opportunity and work my ass off to have stuff. This would impress my wife, family, peers, and friends. I realized that this job, even if I had a college degree or two, I'd be a slave if it were not my passion. The boss man most likely would never have allowed me to do what I love, and this alone would slowly eat me alive.

A portion of my earnings would be taken out of my check to pay for this magical health insurance, which would cause me stress, which is the 4th leading cause of death. They say that even if I feel healthy, I must periodically use this mandatory insurance and visit a doctor approved by my plan for a check-up. This is so he can find something to treat me for, and his treatment could also kill me. By the way, I've never been to a doctor who offered me drugs, even if I was healthy. He'd say, "Take it only as a preventative."

So, I'd get to be a slave for this job and may also need to get jabbed with an unproven inorganic foreign substance to keep me well, but that could make me sick or even kill me dead once again. Are you aware that some life (death) insurance policies state that the policy is null and void if you used experimental drugs like the type used in these jabs, even if urged by the job, government, or your doctor?

The average citizen is worried sick about rising crime, high taxes, and escalating bills; they vote for protection, choosing the lesser of two evils. Fearful, they stay glued to the Tel-lie-vision or social media, buy security system cameras, arm themselves, install digital locks, carry tasers and pepper spray, take self-defense classes, hide their passwords, and are afraid of the dark. Meanwhile, they are unaware of the ac-tual crime that occurs in broad daylight. In most cases, people signed up unaware of the outcomes that eat them alive.

It seems to me that greedy and parasitic billionaire men slyly manipulate the emotions of the masses using auto-suggestive ideas, subliminal messages, hypnotic phrases, and carefully crafted insinuations. These mental triggers convey doom, gloom, fear, and terror and amplify our survival instincts. The people who do this are masters of illusion and mind-control monsters who force the media to conjure stories, fabricate events, and create deceptive reports that play with people's emotions.

Like plastic puppets, the masses unwittingly react, cry out, and demand solutions to the terrible, life-threatening problems the evil manipulators created. These creatures convince their prey that they know what's best for their safety. They coldly craft solutions and devise scenarios to cure the problems they create, which, unaware, willing people pay dearly for.

For a spell to work this well for over 100 years, it must be re-inforced every second day and night. The media must be bought and paid for while being forced to manipulate every emotion to ensure that the sleeping people never awake and find courage in mass.

Imagine a group of sly wolves eating the farmers and putting on peaceful-looking human masks. Then, they proceed down to the chicken coop, sheep pens, and cattle yards, showing pictures of free, happy animals in nature. They also use words, symbols, and sounds that convey compassion, liberty, and peace. These actions trick the already-stressed, tiny brains of caged, domesticated animals into believing that the sly wolves have come to free them. What do you think happens next?

In board rooms, these reptilian-minded, filthy, rich, parasitic people get fatter and fatter by eating us alive a little day in and day out in the name of love, law, order, intellectualism, and

God. Their message is that they know what is best for us all when it's only about what's best for them.

By the way, these same techniques worked so well that marketing firms of all kinds successfully employ them, using fear to get people to buy into almost anything, whether it makes sense or not.

I have an extensive family and have had lots of clients who have passed away at the hands of a system that was supposed to heal them. As I mentioned earlier, Iatrogenesis, or medical mistakes, is the 3rd leading cause of death in the Western world. A great book on this subject is Confessions of a Medical Heretic by Robert S. Mendelson, M.D. Open it to any page, and you'll learn much about this Western Medical Business. I could go on and on about my experiences inside the old-school Western Medicine Machine. Here are a few examples.

My father was told he had diabetes and was prescribed Insulin for 25 years. After more than a year of discussions and arguments with Dad, my mother and I finally convinced him to see a new doctor. It was one of the most challenging things I've accomplished because my father seemed to feel that his old MD was his friend, and he didn't want to upset this guy.

We found a young MD and a naturopathic physician. He took one look at my father, checked the pulse in his arms and legs, and said, "You don't have diabetes; you have a type of insulin poisoning and parasites." He suggested a macrobiotic diet and for him to take my parasite cleanse program for three months, which he did. For the first time in over 25 years, my father no longer had to jab himself with insulin, and the diabetes symptoms disappeared.

The new doctor explained that when he first saw the old doctor in the 1960s, his blood sugar may have been slightly elevated due to overconsumption of domesticated animal flesh and processed sugar. The fat and grease from the domesticated meat had clogged his pancreas, raising his blood sugar levels. My father's mind was blown, and soon, he looked like a new man. My father's insulin habit was costing $2000 a month. So, it cost over $600,000.00 for a medicine he never needed. It also costs him his heart because the overuse of insulin may have another effect called heart disease. After a while, my father attempted to see his old doctor, but the doctor passed away from diabetes complications, just like his father before him.

My mother was diagnosed with Multiple Sclerosis. Her doctors were always giving her experimental meds to try. Most of them had nasty so-called side effects. After hundreds of thousands of dollars changing hands over 30 years, not one of my mother's old-school doctors ever looked at her teeth. After researching mercury, I decided to look at her teeth. She had many old silver-mercury amalgam fillings. Mercury toxicity has symptoms much like MS. New studies are showing that Multiple Sclerosis has been linked to systemic parasites that cause plaque, caused by a combination of heavy metals, excitotoxins, and systemic parasites.

My grandmother was up in age and, besides arthritis, was pretty healthy otherwise. One day, she complained about some chest pains, so they rushed her to the hospital. They said they needed to check her heart and wanted to look inside with some new technology. She didn't like the idea, but "they" convinced her it was best. They inserted a long needle in her neck, and she died immediately. She may have just been having intestinal gas, which is known to cause chest pains. No one asked her what she had been eating.

I remember one specific time I went to a medical doctor once for a check-up, and before he even checked me, he handed me three samples: one antibiotic, one for sleep, and the other for pain. I asked him what these were for, and he said that I might need them based on my chart and my family history. He never looked me in the eyes, which meant to me that he was being influenced to hand them out. The pills did look attractive and colorful, just like candy.

We are finally beginning to witness Western and natural medicine joining forces to create the more balanced healthcare system we need.

There are some outstanding medical doctors out there, and I'd get several opinions before consenting to certain procedures, especially if they are in no way an emergency. It may be advantageous to leave the country to search for alternative treatments if you can. Please do your objective research and consider natural therapies, plant-based medicines, and nutrition. Old-fashioned healing techniques often address the root causes of certain illnesses rather than mask symptoms.

Let me repeat, I feel that modern medicine has its place, especially in emergency cases and disease diagnosis. A medical doctor once told me that "most doctors may not study parasites or nutrition." Therefore, they may not realize or care that parasites and nutritional deficiencies may cause many illnesses. He told me that exercise, proper hydration, supplementation, rest, periodic parasite cleansing, and detoxification are the keys to good health. After he purchased two of my Totalbody Paracleanse programs for himself and his wife, he thanked me for my work, said he supports natural therapies, and asked that I not mention his name.

Man has made significant technological and scientific advancements, especially in emergency care. At the same time, most folks have left behind "age-old" conventional wisdom. Looking at our grandparents and ancestors, we find that people were healthier when they used a more natural approach to wellness.

Does the modern healthcare community care about people's health, or is it more about sick care? The USA is said to be the wealthiest country, but unlike many less affluent countries, it does not have a low-cost or free quality health care system. Why is this?

Keep in mind that medical system mistakes, misdiagnoses, mistreatments, misjudgments, drug side effects, overdoses, underdoses, incompliance, racism, sexism, staff stress, Doctor burnout and stress, overworked, underpaid staff, ignorance, negligence, ego, overconfidence, and lack of empathy are the third leading cause of death in the good old USA?

Don't get me wrong. I am in no way against the good medical doctors and nurses, as there are many. I have dealt with several with good results. Some are friends and clients who are good at what they do. Modern medical care is very good at emergency care. Herbs, for the most part, are used to maintain and support healing. If you have an emergency, call 911, and do not call me if you are looking for an emergency herb.

I'm only pointing out that treating symptoms rather than getting to the root of the problem is a parasitic approach, and it sucks the life force out of people. The Western medical system will eat you alive unless you are awake and aware and get several opinions from other doctors.

Chapter 10:
Germ Warfare

We were told that the first Europeans who came to the Americas sought opportunities to gain wealth and to flee religious persecution. Some were criminals, ex-prisoners, servants, and European slaves looking for a better life. The big piece they left out of the stories told in school was that many of these people were carriers of deadly diseases. They may have had no idea that the pathogens they carried would be fatal to the healthy and free tribes of the Americas.

The nineteen or more weeks of the journey were like hell, and up to fifty percent of them had never survived the trip. They were packed in the cargo holds with farm animals, rats, roaches, cats, dogs, flies, and maggots. The cargo hold floors were covered with excrement, vomit, blood, rotting food, and dead bodies. The people suffered from mouth rot, measles, scurvy, seasickness, influenza, dysentery, boils, gangrene, leprosy, typhus, cholera, malaria, tuberculosis, chicken pox, diphtheria, whooping cough, and syphilis. It is not that all these diseases originated in Europe; they may have never been concentrated on boats this way until European trans-Atlantic journeys began.

It was easy for a few settlers to conquer entire continents of people once the diseases they carried began infecting the native inhabitants. They had no natural resistance to these foreign, deadly pathogens.

In school, we were taught that European invaders and their mighty armies conquered the Indigenous Americans through war. Their victory was due to the diseases caused by parasites that the Europeans unknowingly introduced. Pathogenesis did most of the conquering; this is the beginning of germ warfare. The indigenous people of the Americas had no immunity against these parasitic pathogens, which killed almost everyone almost overnight. As the original inhabitants of the Americas died, so did thousands of years of natural healing arts, information on plant medicines, and proven ideas about living in harmony with Mother Earth, lost forever.

Europeans lived and traveled in close quarters with animals, unlike most indigenous populations of the Earth. The animals may not have shown noticeable symptoms of diseases, but were carriers of zoonotic parasites, meaning they could easily jump from animals to humans. Is this the reason why Europe has had so many plagues? Many scientists say so. Bugs in reality conquered some of the greatest battles in the Americas, not so much by great armies.

Here are two excellent references for this historical information. One is Jared Diamond's book-turned-documentary, *"Guns, Germs, and Steel,"* available from National Geographic. The second is *"A People's History of the United States"* by Howard Zinn.

Check out these facts:

The Spanish flu killed more people than World War I.

The Black Plague, or Black Death, was caused by the bacterium Yersinia pestis, which killed up to 200 million people in Europe. History tells the story of how it was introduced to Europe by Africans, Asians, and people from the Mediterranean. However, there is evidence that poor living conditions, inadequate sanitation, parasites, microbes, rats, flies, and fleas contributed to it.

It's funny how some people need to blame their problems on others especially those they don't know or understand, or those they fear. This is called Xenophobia, an intense or irrational dislike, hate, or fear of people unlike oneself. Now, that is a disease all to itself. Mental spiritual parasites like Wetiko may be at the root of this condition.

Sugar And The Domino Effect

The use of sugar cane can be traced back about 2500 years to what is now considered Africa, India, and Asia. It was eaten raw, juiced, and dehydrated. Like most natural foods, it contains many minerals and enzymes and enters the bloodstream gradually. Pure cane sugar is nothing like the highly processed, refined white cane sugar. Are you aware that processed sugar is the number one drug on the planet, and it inhibits the proper functioning of the brain?

We were taught in school that in 1492, Columbus and others stole the land, enslaved, and killed the indigenous people of the Caribbean islands and North and South America. There are whole tribes of people that are now extinct because of these sick, evil men and their insane lust for gold. Columbus wrote in his log that the Indians were "Fit to be ordered about and made to work, and they ought to be good servants of good intelligence." It's incredible that, even though we now know the facts, Columbus, a criminal, is celebrated with a holiday and revered as a hero.

Around 1503, King Ferdinand, Queen Isabella, and Romanus Pontifex, under the Catholic Pope Nicolas V (Head of the Christian church), sanctioned slavery, and the transatlantic slave industry began. These people said God sanctioned them to do this by proclaiming that the peaceful indigenous people were evil savages. These conquering oppressors were in the business of stealing, possessing, raping, and domesticating people just like cattle.

This was fueled by extreme narcissism, greed, ego, anger, fear, ignorance, and schizophrenia. They may have been infected by parasites affecting the mind, like Rabies, Toxoplasma, Malaria, and an illness some indigenous people call Wetiko. Mentally ill people often profess that God is talking to them when entities possess them. These sick beings started a domino effect that unleashed a physical, mental, and spiritual plague on the world.

This insanity affected their people and their offspring just as much as those they subjugated. The effects still affect all people in one way or another to this day, causing CPTSD, Complex Post Traumatic Stress Syndrome, for the offspring of subjugated indigenous people. This, along with the racism, caused Post Traumatic Slavery Syndrome. These two syndromes created a domino effect that has crippled the advancement of everyone, regardless of background.

I am in no way saying that these few European leaders were the first to take land, kill, and enslave. They were the first to take their mission on the road and overseas, trying to conquer and occupy the entire planet.

The slave trade was not initially about cotton. It was more about the money made on sugar cane, known as "White Gold," as it was costly back then. There was a great hunger for this sweet product in this bitter New World. From the 1740s until the 1820s, sugar cane was Britain's most valuable commodity. Ninety percent of the sugar produced by Europeans was a result of labor.

Cutting and processing sugar cane with its razor-sharp leaves was a job so labor-intensive and hazardous that no one wanted to do it, so slaves and indentured servants were forced to do it. Diseases like yellow fever, smallpox, and malaria sickened and killed most of the European indentured servants and indigenous American slaves. The African slaves were more resistant to illness, especially malaria, so they were forced to become the dominant source of free labor.

It was highly processed, denatured, and refined using Bone Char to make sugar dissolve and look more pleasing to the eye. Bone Char, also known as Bone Black, was made from charred animal bones. The sugar barons have denied this, but it has been alleged that the bones of African slaves were also used to make bone char at times. The sugar manufacturing process creates toxic dust that is highly volatile and often would explode like a bomb, burning anyone in the area and creating more charred bones. Allegedly, on some farms, the cane fields were intentionally set on fire to make the cane easier to harvest by burning off the sharp leaves. Allegedly, sometimes slaves would be trapped in the fields and burned alive, supplying even more charred bones. Who do you think were the bone collectors? There was also a steady supply of African bones from those who died from overwork, punishment, and torture.

Like many businesses, the livestock industry has a zero-waste policy; it uses everything. What do you think was done with all these bones? Dead or alive, the Africans were involved in forced labor. Damn!

This new type of highly refined unnatural sugar has small co-caine-type molecules and is highly addictive. When it touches the tongue, it quickly enters the bloodstream, creating a domino effect. The pancreas jumps into emergency mode and secretes high levels of insulin to counteract the glucose spike and save us. Without insulin, high amounts of glucose can cause diabetic shock and even death. Sugar overload acts like a poison, affecting all organs, especially the brain and the endocrine system (glands). Glands are like minicomputer networks that secrete hormones, chemical messengers. They relay messages from the brain to every cell. Proper brain, glands, and hormonal functioning are essential for the body to operate correctly.

In a docuseries, *THE FOOD THAT MADE AMERICA*, they discuss how, in the late 1800s and early 1900s, people began to leave farms and small towns in mass for the big cities. They hoped to find work in new factories, thinking it would be a better life for them. In most cases, life in these cities was like hell compared to what they had left behind. The food in these cities was terrible because there was no refrigeration yet, much of it was rotten, and there was little to no sanitation. The streets were lined with makeshift markets, tables of deteriorating meat, and rotting vegetables. People were never sure what they were purchasing, how old it was, what it was, or where it came from. Disease was everywhere; people suffered from parasites, fungi, bacterial infections, and viruses. Most people had some illness, and many had stomach cancer.

Where there is a crisis, there is opportunity. During this same period, the pharmaceutical business was being created, and many people were willing to be guinea pigs for just a few cents. If the food and disease didn't kill you, then the crude experimental drugs surely would. Most people still trusted natural cures over these new chemical drugs at this time, but the natural plant medicines they grew up on may not have been available to most living in these big cities.

New food and pharmaceutical companies popped up everywhere; competition was unbridled and ruthless. The unscrupulous salesman's occupation involved manipulating people's emotions to override their natural reasoning through lies, cognitive dissonance, and the art of repetition. This caused people unaware of their own power to become preoccupied with fear, lack, and limitation, real or imagined, which is one of the most powerful tools in sales. In the late 1800s and early 1900s, many new companies were advertising agencies that used any means necessary to hijack minds. First, they would create a product, an idea, a drug, or a treatment. Second, they would make the concept of a need for, or a disease caused by, this product. Next, these wordsmiths used the powers of language, spelling, autosuggestion, and hypnotism to entrap and occupy people's minds, creating pure consumers.

Often, when a family member, comrade, friend, religious or political group, or any authority figure buys into an idea as if by magic, people around them will join in without proof of the idea's validity. People can become like puppets on autopilot, convinced that what they are being told and sold is essential for survival. Whether it made sense, people allowed themselves to be herded like sheep while throwing logic to the wind. This was the birth of a new type of slavery, mental slavery, and mind wars. The salesmen worked street corners as pushers, trying to convince people that chemicals were superior. Sound familiar?

Like with dominoes all lined up, one domino pushed could knock over all the others with little effort. This is where we get the term "pushover."

During the Industrial Revolution, as now, several varieties of crude sauces, gravies, and garnishes were used to mask the appearance, smell, and taste of rotten meat and old vegetables. One never knew what was in these products; they often contained toxic chemicals like formaldehyde and were as bad as, or even worse than, the toxins they were supposed to cover up. Companies popped up everywhere. Ego, greed, and mental illness led businesses and families to war against each other, competing to be "on top" no matter who got screwed. Their only goal was total conquest of everyone's minds, bodies, and pockets. Food companies began using refined sugar, dyes, and rock salt to make food taste and look better, at a time when much of the food in cities was of poor quality. It was the first time this ultra-refined white sugar was introduced on a large scale, and it was a game-changer for these new companies. Before this, natural sweeteners such as old cane sugar, dates, honey, and dried fruits were expensive and sometimes unavailable.

A Medical doctor told me that refined sugar is the number one drug, very close to Cocaine, and people quickly become hooked, making it challenging to get off it. It is cheap, and nearly everyone craves it. Companies began adding candy coatings on pills and sugar to medical syrups. This is where the saying "A spoonful of sugar makes the medicine (poison) go down." Technically, this highly refined product is a drug, and America is hooked on it.

John Stith Pemberton was attempting to break his heroin habit and combined Cola nuts and Cocaine to come up with what would become the most popular drink besides coffee. It was Asa Griggs Candler who purchased it and eventually replaced the cocaine with caffeine and refined sugar. Today, the average 20-ounce can of soda has approximately 14 teaspoons of refined sugar. A major company boasts that more than 1.9 billion of its drinks are consumed daily in more than 200 countries.

The TV series "The Food That Made America" clearly shows this battle of supremacy. The average person's mind, body, and spirit became consumed by a small group of greedy men and one woman. They used the art of manipulation to turn people into consumer addicts hooked on junk. It is an ancient thought that whoever prepares food has a massive effect on those who consume it. What do you think happens when a society eats products created unnaturally by greedy, ruthless corporations and made by workers working under stressful and unhealthy conditions to make ends meet? By the way, the endings never meet.

The primary purpose of the digestive tract is to break down what we eat into glucose, our energy source. There's sugar in almost everything we eat. The digestive tract slowly breaks it down, feeding every cell in the body. The brain uses more glucose than virtually any other organ, which it uses as fuel to regulate and process all body functions. The human brain is a complex organ that does billions of computations each second. It has more processing power than all the computers on earth linked together, and it works 24 hours a day, never taking a break.

Processed refined sugar has a glycemic index of 99, compared to date sugar, which has a GI of about 47, and raw sugar cane, which has a GI of 65. The glycemic index number depends on how fast the sugar enters the bloodstream. Refined sugar causes a sugar spike because it hits the bloodstream when it touches the tongue. This type of sugar affects all parts of the brain, especially the amygdala, hypothalamus, and pineal glands, and inhibits their natural functioning.

The hypothalamus is the master gland that regulates the entire endocrine system and links it to the nervous system. It regulates temperature, thirst, hunger, sleep, and emotions, and balances all body systems. Proper functioning is essential for physical, mental, and emotional health. When we ingest enough glucose to fuel the body, fat cells produce leptin, a hormone. Leptin signals the hypothalamus that we've had enough glucose or anything else. The hypothalamus signals to the nervous system that we are satisfied and should stop consuming.

Processed refined sugar causes spikes that overload the entire brain with glucose. The hypothalamus becomes paralyzed, and the nervous system cannot tell whether we've had enough sugar, nor can it create homeostasis in the body. This causes us to desire more sugar, food, and everything. This condition profoundly affects us emotionally, and we feel no joy or satisfaction. At this point, a person becomes hooked on trying to feel good but never really can get or stay there for long, just like a person hooked on any other drug. This is where the term "On Dope" comes from because the hormone dopamine goes into overdrive. America and much of the world have become consumed with consuming.

The pineal gland is our spiritual antenna that collects information from our environment and is responsible for intuition, creativity, and individuality. Processed refined sugar shuts down the pineal gland, disconnecting us from self-awareness and elevating us above most other animals. The Western world became hooked on sugar and lies, and people lost their own innate sense of spirituality, culture, truth, family, and love. It became easy for the human parasites to convince the average person to let go of their natural spirituality and believe that they needed the church or an old bearded white man in the sky to guide them. People became lost, and the natural human colonies began to collapse. People lost the ability to think logically and be discerning, and they could no longer make choices for themselves. They became easily led, controlled, manipulated, taken advantage of, and domesticated like barnyard animals, too afraid to break free.

The bitter truth is this sweet toxin has thousands of detrimental effects on the body and mind, ultimately causing organ failure and insanity.

This created a windfall for the drug companies, who promised to heal the sick by giving them more and more poison, creating more drug addicts. In the late 1800s and early 1900s, a handful of greedy business tycoons became human bloodsucking parasites, and 80% of the people became their hosts. Parasites crave processed sugar because it weakens the host's natural defenses. When the host is high in sugar and other toxins, there's less chance of detection. Sugar helps parasites cloak themselves and provides the energy they need to survive.

A medical doctor once told me that people hooked on cocaine and those caught on processed sugar shared a roundworm called Ascaris lumbricoides. The parasite lives in the digestive tract, disrupting the gut microbiome, attracting Candida and other parasites, and creating a swamp. As the parasite takes over the host's thinking factors, the host craves more and more glucose, often in the form of alcohol, wheat, and gluten.

Wheat is one of the most hybridized, crossbred, and cloned plants ever. Hard Red Winter Wheat is commonly used in the USA. It is higher in protein and therefore has higher levels of gluten.

Soft White Winter Wheat is commonly used in Europe and the eastern regions. It has less protein and, therefore, less gluten and tastes better. Is this why people in the USA have more wheat-related health challenges than people in Europe, Asia, and Africa, even though they eat just as much wheat?

Wheat contains gluteomorphin, an opioid or morphine component of gluten protein known to cause addiction. Opioid substances turn the receptors responsible for feeling things like pain while inhibiting and manipulating all emotions. This combination creates a state in which the host no longer has natural gut feelings or intuition. The gut has more neurons than the brain and can be considered the first brain.

The gut naturally sends more messages to the brain via the Vagus nerve than the brain sends to the gut. At this point, the host is caught in a feedback loop between Toxoplasma, Ascaris lumbricoides, Candida, processed sugar, and the standard American diet. When a person craves sweets, is always hungry, is chronically depressed, spiritually lost, forgetful, never satisfied, and lives in fear, they are most likely host to several types of parasites. This is the condition of 70% or more of the people, and most are unable to get off the bittersweet ride; they are hooked until they die.

Alzheimer's Dementia

More and more information is coming out about Alzheimer's, aka Type 3 Diabetes, and its connection to processed sugars. Many of our clients are witnessing great results when they do our detox programs for 3 months, cut out processed sugar, gluten, and high lectin foods. They have also followed a supplement protocol with Lithium (orotate), Citicoline, CoQ10 with PQQ, and a Nicotine 7mg Patch. Yes, you heard me, Nicotine! Remember, they lie and benefit when the wise elders forget who they are.

Processed sugar, drugs, and parasites have domesticated and enslaved the Western world, consumed by "The Domino Effect."

Chapter 11:
Vital Nutrients

Parasites and toxins damage and suppress the immune system while robbing the host of vital nutrients. Strategic supplementation is essential to revitalize the body during cleansing and beyond.

Vitamin A: Immune Signal & Tissue Repair
Certain parasite larvae crave vitamin A and will siphon it from the host. Vitamin A is crucial for healthy immune signaling and epithelial integrity. When parasites drain your reserves, your defenses drop and they survive.

Common signs of low Vitamin A: frequent colds and flu, skin and respiratory issues, night blindness, reproductive problems, slow wound healing, dry skin, dry eyes, and what I call **"dry personality syndrome."**

Rebuild while detoxing: emphasize seeds, nuts, sweet potatoes, yams, leafy greens, and winter squash to replenish vitamin A precursors naturally.

The mineral Boron is essential for helping the body fight parasites, regulate chronic yeast infections and other pathogens, address estrogen dominance and bone thinning, prevent premature aging and testosterone decline in men over 40, support immune function, address chronic inflammation, support menopausal symptoms, support a healthy brain, nerves, and regulate emotions. Several foods naturally contain boron, but because of depleted soil minerals, I would research a natural supplement.

Vitamin B12 & the Gut Terrain
Fish tapeworms can absorb Vitamin B12 from their host. Once levels are low, it can take **six months to a year** to re-establish optimal B12 status. B12 supports healthy stomach function, including the production of hydrochloric acid, which supports digestion. When B12 and gastric tone are low, pockets of poorly digested food in the gut (diverticula) become parasite playgrounds.

Vegetarians and vegans are especially prone to low B12 and may benefit from targeted supplementation. A B-complex supports new cell development, blood formation, digestion, and healthy brain and nerve cells, which are vital during and after detox. If you wrestle with low energy, brain fog, or digestive complaints, B12 deficiency may be a contributor.

Zinc + Vitamin D3 (Soltriol): Immune Backbone
Zinc, alongside vitamins A and D3, is fundamental for immune resilience.

Vitamin D3 (Soltriol): Once referred to as *soltriolsol* meaning light vitamin D3 behaves more like a hormone than a vitamin. It's a **solar transducer** and messenger a somatotropic activator that helps regulate systems involved in growth, repair, and performance. Your skin manufactures D3 (calcitriol) when exposed to sunlight.

Why it matters: robust immunity; strong bones and teeth; clearer thinking; steadier moods; cardiovascular and lung support; insulin dynamics; courage and critical thinking. People living far from the equator need more Vitamin D3, especially in winter. Deficiency is linked with **Seasonal Affective Disorder (SAD)** depression, stress, bipolar-like swings, chronic fear, and identity drift.

Most people in North and South America benefit from more sunlight, D3-rich foods, and a **well-made D3 supplement.** (Not all "natural" supplements are equal; I favor **Jarrow Formulas** for quality.)

Melanin, Sunlight & Bioelectric Vitality

Humans are naturally **photoelectric and** we convert sunlight into biological energy, much like solar panels. **Darker skin can safely absorb more sunlight**, translating to greater life-force potential when sunlight is adequate. The more life-force you house, the harder it is for certain parasites to set up shop.

Melanin is a **built-in sunscreen** that helps protect against harmful UV while allowing beneficial light to in. Sunlight exposure also helps the pineal gland produce **melatonin** a master antioxidant and hormone linked to sleep, memory, focus, will, and intuition. A high-functioning pineal gland and sufficient melatonin make it harder for mental and energetic parasites to take hold.

A physician once told me that colonizers forced clothing and shoes on indigenous peoples to **break their souls,** dull their natural spirits, and disrupt tribal unity. That's Deep. The less we block nature, the healthier and more spiritually connected we tend to be. Consider: unless you live near the equator, sunlight travels a longer path to reach you, especially in North

America, so highly melanated people often need deliberate sun exposure and/or D3 supplementation to stay spiritually, emotionally, and immunologically balanced.

Foundational Mineralizers: Humic Acid & Shilajit

They energize melanin, help chelate heavy metals, nourish the pineal gland, and deliver a near-complete spectrum of trace nutrients. I combine them in **Doctah B's Platinum Life Monatomic Elixir** to support intuition, healthy aging, telomere integrity, heavy-metal detox, environmental protection, pineal enhancement, peak performance, core energy, and increased health span.

I later crafted special variations of the **Beckwith Reserve** and the **Gregory Reserve** for Rev. Michael Beckwith (Agape International) and the late Dick Gregory.

Black Walnut Oil

It offers a balanced profile of essential fatty acids (3, 6, and 9) to support whole-body vitality. It also nourishes melanin dynamics and healthy hormone balance. It also helps maintain healthy melanin in skin cells.

Chaga mushroom

It's a powerhouse in the medicinal mushroom world, helps **repair and support melanin** while delivering exceptional antioxidant capacity. Chaga is one of the 11 mushrooms in my **Master Mushroom Tonic**, formulated for immune intelligence, intracellular rejuvenation, and deep antioxidant protection.

Healthy Fats: Coconut, Avocado, Olive

Unrefined Coconut Oil brings omega-rich fats with vitamins A, E, and D3. It supports digestion, brain nourishment, and immune function, and exhibits antiviral, antibacterial, and antifungal properties. Ignore the smear campaigns that mostly came from competing seed-oil industries. Your body uses coconut oil to repair tissues and support healthy hormone and nerve function. Heat tolerance: ~350°F. Great for light cooking, smoothies (1 tbsp), moisturizing skin, soothing dry eyes, calming an upset stomach, and even as a gentle sun-tanning oil.

Avocado oil tolerates higher heat up to ~485°F and, like coconut oil, offers a broad spectrum of healing fats for cooking and finishing.

Moroccan olive oil is world-class. A staple across the Mediterranean, where people routinely enjoy better cardiometabolic health, mobility, and longevity.

Vitamin C: The Antioxidant Conductor

Vitamin C deficiency is among the most common on earth. Vitamin C is essential for collagen formation, immune function, and the absorption of key minerals (notably iron). It protects genetic material from oxidative "rust."

Humans **don't** manufacture Vitamin C; we must get it from food and supplements. Many commercial formulas are **acidic**, which can clash with already acid-forming diets and stress chemistry. That's why I created **Doctah B's Vitamin C Tonic**a more **alkaline, food-based** blend featuring:

- **Camu camu** (Brazil)

- **Baobab** ("tree of life," Africa)

- **Sumac berries** (Middle East)

- **Organic citrus**: oranges, bitter orange, lemon, lime, grapefruit

This combination delivers high levels of naturally buffered Vitamin C, along with protein and polyphenols. It's adaptogenic tonic, safe, and bioavailable compared to many synthetic acidic forms. Baobab offers minerals, micronutrients, and prebiotic fibers for gut ecology.

Camu camu is exceptionally rich, often cited as having far more Vitamin C than most fruits, plus flavonoids and ellagic acid. It's anti-inflammatory, mood-supportive, and antimicrobial.
For added immune intelligence, the tonic includes **colloidal zinc and copper** in harmony.

Keep in mind

- **Replete what parasites deplete** (Vitamins A, B12, D3; zinc; Vitamin C).

- **Fortify the biofield** (humic acid, shilajit, chaga, black walnut oil).

- **Feed the nervous and hormonal systems** with clean, heat-appropriate fats (coconut, avocado, Moroccan olive).

- **Sunbathe often** and, when needed, supplement D3 so the pineal, melanin, and immune orchestra can play in tune.

Revitalize the terrain, and what feeds on you can't thrive in you.

Magnesium is a master mineral that regulates 300 enzyme activities. It is one of the top deficiencies, especially in those infected with parasite pathogens. Magnesium is essential to balancing hormones and the nervous system. It also helps regulate blood pressure, insulin resistance, protein creation, energy, gene repair, and anti-inflammatory responses. It can also help with depression and stress reduction.

Dark chocolate, pumpkin seeds, black beans, quinoa, and avocado are high in magnesium. There are many Magnesium supplements, but I like magnesium chloride, magnesium glycinate, and magnesium taurate, which are more absorbable and support the circulatory system.

We use ancient trace minerals from the Great Barrier Reef as a base in our Tonics and Elixirs. These minerals are high in magnesium, calcium, potassium, and sodium, along with all the minerals in balanced proportions present in the most pristine seawater.

If you find that you are experiencing frequent muscle cramps, high blood pressure, stress, migraines, and sexual dysfunction, then this may be a signal to increase your intake of Magnesium.

Cabbage, Asparagus, and Radishes contain substantial amounts of bismuth, which supports a healthy digestive tract. Bismuth helps repair damage caused by the superbug H. Pylori, which, if left unchecked, is a common cause of ulcers and other chronic illnesses. Note that cabbage alone cannot prevent or eradicate H. Pylori; a special protocol is required.

The US government felt that Vitamin B17 posed a threat to public safety and made it illegal as a supplement. Bitter Apricot kernels possess high amounts of Vitamin B17. Apricot kernels contain Vitamin B17, also called amygdalin, and Garbanzo beans contain a small amount. Naturalists have traditionally used B17 for cancer prevention and treatment. The mainstream medical community says there is no proof of this claim, and no human trials have been conducted. Interestingly, indigenous people have used natural substances effectively for thousands of years, but that is not considered proof. It looks to me like indigenous people are not thought of as human beings. Before you get all upset, maybe you should look up the definitions of human, civilized, citizen, and, while you're at it, research the meaning of domesticated.

Google AI says the word 'human' has a complex etymology, tracing back to Proto-Indo-European (PIE) roots and evolving through various languages. I found something interesting about this in an old Blacks Law Dictionary. English is a tricky language. People are becoming human doings vs beings.

One day, I realized that I am not a human being, and the idea of trying to be like one has been eating me alive for a long fucking time. There, I said it. Wow, I feel much better now!

Ok, back to the vital nutrients.

I would ask people to research Vitamins B17 and B15 and the truth about nicotine. You will find their history very interesting.

Excitotoxins are highly toxic chemicals used in food processing. They destroy brain tissue and the nervous system and are partly, if not wholly, responsible for several mental conditions. Excitotoxins overstimulate brain and nerve cells, causing them to fire so rapidly that they die prematurely, and the FDA approves them.

Excitotoxins are in many foods in major stores, and there are over 60 different names for them. They are used as flavor enhancers and preservatives. These poisons are in so many processed foods. It is challenging to find foods without them. Babies get a total dose of these brain-frying toxins in some baby foods fed to babies in the name of love by their mom. Could this be why mental imbalances are the norm in children nowadays? We do offer an Excitotoxin Relief kit, which helps reverse excitotoxin damage.

One of the most common excitotoxins is the flavor enhancer MSG. It artificially intensifies the flavor of foods, making people crave more of whatever it's put in. It also makes us resistant to Leptin, the hormone that signals satiety. Leptin usually signals the hypothalamus gland that we have had enough of something, especially sweets, which feed parasites.

The hypothalamus is the interface between the endocrine (glandular) system and the nervous system. Its primary function is to maintain homeostasis, or balance, throughout the body and mind.

MSG intensifies the tastiness of any treat, making you desire it even more. Second, MSG has been shown to make us resistant to leptin (the hormone that makes you feel full). Why would you ever put down a snack if your brain never gets the message to stop eating it? Finally, MSG causes the secretion of insulin, your fat-storage hormone, which drops your blood sugar and makes you hungrier faster.

Here is a list of names for MSG from Prevention.com

- Autolyzed yeast
- Autolyzed yeast protein
- Calcium glutamate
- Carrageenan
- Glutamate
- Glutamic acid
- Hydrolyzed corn
- Ingredients listed as hydrolyzed, protein-fortified, ultra-pasteurized, fermented, or enzyme-modified
- Magnesium glutamate
- Monoammonium glutamate
- Monopotassium glutamate
- Natural flavors (ask manufacturers for their sources to be safe)
- Nutritional Yeast
- Pectin
- Sodium caseinate
- Soy isolate
- Soy sauce
- Textured protein
- Vegetable extract
- Yeast extract
- Yeast food

Cannabis

Cannabis and hemp are among the most amazing plants known. While the number of products that can be made from them is impressive, the ways they support a healthy planet and a strong economy are staggering. The list of documented medicinal uses grows each day. One of the main things this natural plant family addresses very well is stress and inflammation, both leading causes of physical and mental illness. They both support the body's endocannabinoid system (ECS). This system helps maintain homeostasis, or balance, throughout the body by connecting all organ systems like a hyper-speed Internet service. It is a network of receptors, molecules, lipids, and enzymes that facilitates the transfer of information between organ systems. This system works much more efficiently than modern humans overworked, nervous, and immune systems.

We are physically and mentally healthy when communication between body systems is faster.

One of the thousands of components of the ECS system is the potent mood enhancer anandamide. Ananda is a Sanskrit word meaning bliss, joy, happiness, or pure pleasure. This component effectively reduces the stress response, boosts the body's natural defense against illness, and speeds up recovery time. This is important because stress is literally eating away at the population.

The cannabis family contains THC and CBD, which are the fuel for the ESC system. Why were they demonized for so many years? People who use these plants for mind and body medicine are less likely to be stressed, ill, or easily controlled. Could they replace many pharmaceutical companies and cut into drug companies' profits? This plant has personally helped me stay focused, calm, creative, balanced, and healthy.

Probiotics

We must support our gut bacteria with probiotics, intestinal flora, and friendly bacteria. They create what's called the microbiome an entire universe of microscopic creatures living within each of us. We could in no way live without these fantastic beings. They are responsible for digestion, a large part of our immune system, and significantly affect who we are, our thinking, and our mood. Parasites and Candida disrupt the natural balance of intestinal flora. Even after a 28- or 90-day parasite cleanse, we must constantly reinforce the friendly bacteria in the digestive tract. Without enough friendly bacteria, we cannot truly be free to tap into our true potential or ever be truly healthy.

When the balance is thrown out, we are easily compromised and possessed, physically and mentally. One becomes possessed, depressed, bipolar, and in some cases tri-polar. Other symptoms are acting hyper-reactive, nervous, ADD, ADHD, autism, Parkinson's, diabetes, and being over- or underweight, to name a few. An imbalance in gut bacteria is one of the leading causes of chronic constipation, candida, or yeast overgrowth. And let's not forget the leaky gut syndrome that plagues many who eat the Standard American Diet. Leaky gut syndrome is when undigested food particles and waste material leak through the damaged walls of the gastrointestinal system into the blood. Low probiotic levels, parasites, processed foods, plant lectins, and yeast overgrowth cause this.

This condition leads to chronic inflammation, blood toxicity, mental health issues, cancer, high blood pressure, violence, and a very long list of illnesses unheard of by our ancestors. This is why people no longer have or trust the notions called "gut feelings." This is why so many people are mentally full of shit.

There are more neural connections in the gut than in the brain, and more connections from the gut to the brain than the other way around. The gut is truly the first brain. A healthy microbiome supports willpower and being more instinctive, self-aware, and proactive. An imbalance in the gastrointestinal system makes people easy to control like sheeple (civilized, domesticated, sheep-people). When people become chronic consumers, they can no longer think for themselves or make wise choices to help them thrive. They just barely survive like cattle.

Essential foods that help increase friendly bacteria include fermented foods like pickled beets, sauerkraut (pickled cabbage), olives, and coconut kefir. Our way of life in the Western world has destroyed several essential bacterial strains. This has compromised our health on all levels and supports the health of pathogens of all kinds. We need at least 24 different strains of friendly bacteria to be healthy. Many probiotic brands tout having billions of beneficial bacteria. This may sound appealing from a marketing perspective, but what matters more is the number of strains in the probiotic product. These friendly bacteria also need a specialized source of nutrition and fiber called prebiotics, fructooligosaccharides, or FOS. These are present in coconut meat, yams, green bananas, plantain fruit, yacon fruit, and monk fruit, among others. The last two are great natural sweeteners with complex sugars and are safe even for diabetics because they never enter the bloodstream. Garlic, leeks, artichokes, onions, and asparagus are also excellent sources of prebiotics.

Raw vegetables and fruits contain lots of fiber, which helps to scrub the digestive tract of waste gently and creates a habitat for friendly bacteria to thrive.

Coconut Kefir has at least 30 different strains and helps replenish the microbiome. There are two types of coconut kefir.

One is made from coconut cream, which is thick and white like yogurt. The other is made from young coconut water. These are both excellent sources of essential-friendly bacteria. We must be careful when choosing acidophilus and Bifidus probiotics because they may have been produced using milk.

Cow's Milk: The Secret Weapon

Many highly melanated people whose ancestors originated around the equator don't make lactase after they are weaned off their mother's breast. Lactase is the enzyme that breaks down the lactose in milk. This alone is a significant contributor to widespread illness amongst black and brown people. We have been convinced that cow's milk suits everyone, but the opposite is true for certain ethnic groups. This may be true for some of European descent because they may still produce enough lactase to properly digest raw cow's milk. This fact alone can explain why non-Caucasians who consume dairy from cows lose the ability to tap into their gut mind.

Many studies show that commercially processed cow's milk harms good health and well-being. The dairy industry believes injecting cows with recombinant bovine growth hormone (rBGH) is okay. This is a synthetic, man-made hormone that causes cows to produce more milk while shortening their lives. They figured out how to splice genetically manipulated cow genes into genetically engineered bacteria and reintroduce them to antibiotic-dosed, cooped-up, domesticated cows. If this cocktail shortens the life of cows, what will happen to the people who drink it? How do they keep cows pregnant so that they perpetually produce milk throughout their lives anyway? Why do you think they would do all this? I'll tell you why; it's because the demand for dairy products, including dressings, blue cheese, or ranch, is getting larger every minute of every day. It is also best to abstain from all dairy products because the mad cows are returning to haunt you.

I highly recommend the book *MILK THE DEADLY POISON BY ROBERT COHEN*

It is essential to eat foods that grow in season and not food that comes from thousands of miles away. These foods weaken the body, allowing parasites to thrive. Ever wonder where the winter watermelon comes from? It may have come from thousands of miles away. This is yet another example of things we have been conditioned to accept as normal, but it is totally unnatural.

We must again support local farmers and buy food that grows locally because local foods do not have to travel far, so they retain more life-force energy. Every time you pick a fruit or vegetable off the vine or tree, it begins to lose life force energy each hour. Usually, within 12 hours, a plant cut or a fruit picked may lose 30% or more of its life-force energy. When we get food from farmers, we get it with more life-force energy faster. A lot of people in North America are into food aesthetics. So, if the food looks healthy, they think it is healthy. Yet, sometimes, looks are deceiving. Most of those foods were manipulated, sprayed with toxic pesticides, and polished up to look wholesome. Certain types of pesticides are so bad that they are banned in the USA. This sounds good, but the truth is, the good old USA sends these poisonous pesticides to countries south of the border. Then, it contracts these countries to grow food for the USA using these pesticides. One such food is non-organic bananas. An easy way to identify an organic fruit or vegetable is to look at the small PLU label for a five-digit code that begins with a 9. **94011** is the code for organic produce, and 84011 is for genetically engineered produce. **4011** is the code for standard produce that is most likely treated chemically. Also, beware of the Apeel label, which is said to cause fruits and vegetables to stay fresh longer. How? Look up the ingredients and side effects of consuming this stuff.

The pest control chemicals used over time are making the bugs mutate and become stronger while becoming increasingly toxic to people. I have concluded that someone or something considers people pests to be controlled.

While shopping, consider that foods commercially grown in North America may not contain enough minerals because they are grown in mineral-deficient soil, or even without soil, and often under artificial light. We need at least 90 minerals to be at our best, and these commercial crops may not have enough to promote good health. These plants may still grow and look great, but they cannot support the life force people need to be healthy and break free of the psychological spell most people are under.

If we used more organic and natural ways of raising food, we would have more energy, life force, and vitality, and so would we.

Microwave ovens work by sending high-frequency microwaves through foods, heating up the molecules. These radio waves penetrate glass, paper, porcelain, plastic, and human flesh. This process zaps essential nutrients out of foods and can cause microwave damage to all nearby living things.

I would never use radiation to cook food, even though the government says it's okay, unless my brain has been fried. They say the long-term effects of this nuke box are still being reviewed. Do you think they ever publicly connected major illnesses to using that contraption and discontinued its use? I doubt it.

I implore all people to research where their food comes from. Is it actually coming from a farm or a factory? Who is transporting it? Are chemicals used during the farming and transportation processes?

Neem, also known as the margosa tree, is a natural pesticide. It naturally eliminates harmful insects and parasites in the garden. When used correctly, neem seed oil is absorbed by plants and helps them inhibit the growth of crop-destroying pests.

The many health-supporting qualities and pest-control effects of Neem are well documented and available at garden and farm supply stores. I use neem in my TotalBody Para Cleanse programs, and it has been used effectively in the Ayurvedic healing traditions of India and Africa for thousands of years. Why would neem not be widely used by the industrial farming complex corporations?

Humic and Fulvic Acids were mentioned earlier when I discussed melanin. I'd like to mention them again when raising natural crops. Humic and fulvic acids are created from the decomposition of plant material. They have a high micronutrient content because they contain all known micronutrients. They are both used as natural plant fertilizers and to enhance growth.

I worked with scientist, researcher, and agroecologist Dr. Robert Faust, the top authority on humic and fulvic acids. He introduced me to adding vital nutrients, the energy essence, to natural products as a last step in formulation. During the formulation of natural products, energy is lost. Most bottled natural products may contain only 33% of the energy plants had initially before processing.

Everything changed for me once I learned the protocol for augmenting the energy levels of herbal products. Adding humic and fulvic acid to my formulations took time to master because these elements can cause the end product to ferment quickly, and precise timing is essential. I now include humic and fulvic acids to enhance the healing potential to a quantum level in most of my tonics and elixirs. These products contain numerous micronutrients in a compact form and are therefore considered quantum superfoods. This critical piece in parasite eradication is all about energy augmentation, increasing the biorhythms while alternating frequency over 28 days. This unique process is of utmost importance for outwitting these resourceful and resilient beings known as parasites.

Chapter 12:
Physical Parasite Eradication

Parasites take full advantage of the 28-day moon cycles and thrive when in sync with seasonal changes. They live and die according to these natural rhythms and tune in to the host's eating and living habits. When we change our habits and seasonally detoxify our bodies, we boost our natural defenses and weaken them while disrupting their mating and feeding cycles.

Systemic parasites are more challenging to eradicate from the body than intestinal parasites. For instance, to deal with parasites in the brain, you need products that can pass the blood-brain barrier. Most herbs cannot pass this barrier because they would damage the delicate brain tissue. The key is to affect the energy and chemistry of the blood itself.

I researched and used all-natural products with unique properties that emit specific tones that could change the energetic frequency of the entire human body, especially the blood. This makes the environment within the host less favorable, irritating the parasite's senses and creating conditions that are toxic to its survival while boosting the host's immune system.

From my research, I realized that eradicating systemic parasites would require a 28-day program based on the same moon cycle the parasites used to flourish. After much research and trial and error, I finally created this protocol called

ELEVATED TOTAL BODY PARACLEANSE, which includes adaptogens that tune the body, rare herbs with unique phyto-chemical profiles, trace minerals, essential oils, flower essences, crystal elixirs, photon waves, and sound vibration.

This protocol helps create an uninhabitable environment within our bodies for these parasites. It can disrupt their eating habits, destroy eggs and offspring, and interfere with their communication and navigation systems. Because systemic parasites are masters of survival and adaptation, each season we change the elements of the program to stay ahead of their ability to figure it out and adapt to it. In other words, certain herbs used in this program must be alternated to prevent adaptation and mutation into superbugs. For this reason, completing the program each season may also be advantageous, as some parasites are seasonal.

This may be why the average parasite-detox product is only effective against simple intestinal parasites, and only for a short time, before these intelligent beings outsmart us and return even more potent, with a vengeance. In other words, the average parasite cleanse may kill weak parasites while making the strong ones even stronger.

Over time, we realized that people with a chronic illness experience better results when they follow the ELEVATED TO-TALBODY PARACLEANSE PROGRAM for at least three months, that is, a whole season, ninety days. It takes ninety days to restore balance and to change the rhythms or habits that have caused the imbalances. It is essential to open the body's channels of elimination and maximize our natural defenses. It is not the program that eliminates parasites on its own; it helps create homeostasis and optimize our body's natural ability to heal, raising the body's energy field. This interrupts the parasite's lifecycle.

A program like this may be too intense for children under ten. Also, some research suggests that young children in the Western world may not harbor the same parasites as adults. One traditional way to help detoxify young children from parasites is to make a tea with thyme leaves, papaya seeds, raisins, cloves, black walnut hulls, orange peels, and pomegranate seeds. The goal here is to lure the parasites to the sweetness of the raisins, and the fruits and herbs in this formulation can help eradicate them.

Parasites that inhabit organs can cause many misdiagnosed illnesses. Research has shown that the mainstream chemical approach to eliminating parasites has been only minimally effective. Like most pharmaceuticals, it has undesirable side effects.

Back in the day, almost every time I went to a doctor, they would offer me an antibiotic or some vaccination to keep me from being sick. As an adult, I always said, "No, thank you". I always wondered why they got so angry when I chose to do what Nancy Regan said to do. "Just say no to drugs."

A medical doctor told me that antibiotics are effective only against certain types of bacteria and can destroy beneficial bacteria if not replaced, which may take months or even years. Most parasites are not like single-celled bacteria; antibiotics can benefit them. Antibiotics are used very liberally today. They are fed to livestock to maintain health and promote faster, larger growth. Antibiotics may have no effect when needed because they are already present in the body from consuming treated livestock over time. Yes, commercial meat eaters may unknowingly be taking antibiotics, vaccines, and growth hormones every time they eat factory-farmed animal flesh.

Food can be either your medicine or your poison. If it is packaged, processed, enriched, or genetically modified, it is most likely poisonous. To eradicate the beings eating us, we must now wisely choose what we consume daily. A little bit of poison can be tolerated occasionally. Wise ones know that eating habits can determine who we are and who we are not. Food is information, and "You are what you eat" is accurate garbage in, garbage out.

The masses may be unaware that they are involved in nutritional genocide and suicide. Each person must choose for themselves whether to break the spell of this sick system or continue to be eaten alive.

During detoxification, we should limit or avoid the following foods: simple carbohydrates, such as highly processed foods; American commercial white flour; processed sugar; factory-farmed animal fat or grease; and foods containing high levels of gluten, such as hybrid wheat. These highly processed products create a feeding ground for parasites and harmful microbes. To become and stay healthy, I would avoid fast food at all costs. They are detrimental to all the body systems, distorting objective reasoning. Some natural sweeteners, such as refined honey, processed cane sugar, and commercial fruit juices, must be avoided while detoxing or to stay healthy. Simple sugars are what parasites crave; they feed yeast and create an environment for chronic diseases like cancer. Did your allopathic doctor inform you about this?

Food addictions can be challenging because the parasites, mineral deficiencies, and candida albicans (yeast) overgrowth are at the root of these food cravings. Breaking these bad habits can be achieved through our seasonal parasite cleansing and detoxification. This step-by-step behavior modification, along with the nutritional guide in the booklet that comes with the 28-day cleanse program, can help you take back your life. When we kick the pirates off our vessel, the addictive habits and associated illnesses go with them. Instead of blaming the system, it could be a good time for us to take the wheel now and become the captains of our vessel.

Most parasites are affected by the 29.5-day moon cycle and the 28-day human biorhythm cycle. These circadian rhythms are like a drumbeat or pulse wave, influencing all living things on Earth.

When the body's environment is inhospitable to parasites for at least 28 days, the parasite's eating, mating, and rest cycles are disrupted. This causes the parasite to either die, leave, or enter quiescence or hibernation. This is true for both physical and mental parasites.

As we begin to eat with more awareness, we elevate our entire lives. To become champions, we must eat what supports our higher goals as champions and stop eating like losers. To effectively break the hypnotic spell of parasitic possession, we must repossess our bodies, and this starts with our nutritional choices. What the consumer chooses to consume determines whether they are being consumed or not. Choose wisely.

Chapter 13:
Intra and Extraterrestrials

We are in no way alone in the universe, even though we've been led to believe otherwise. That's exactly what the elites want us to think, so they can keep their illusion of ultimate power and control intact.

There are at least 100 billion stars out there, and countless planets where life has evolved in ways completely different from ours. Some beings might not even be made of carbon or nitrogen like we are. There are worlds made of pure light, vapor, plasma, and maybe even noble gases, realities we're just beginning to imagine.

Humans are still in the infant stages of science, space exploration, and cosmic knowledge. Every year, we stumble upon new evidence of life forms right here on Earth that we never dreamed existed. Even inside the human body, we're finding new organs. We still have no idea how many species exist in the ocean or within us. Modern man knows far less than he pretends to know about life, the cosmos, or even his own planet. You can see it clearly in the way we treat each other and the Earth itself, the home we're supposed to love.

The Dogon of West Africa had deep knowledge of the stars long before Western man had telescopes. They already knew about the Sirius system and its companion stars in incredible detail. To scientists, these people were considered poor, uneducated, and primitive. But the truth is, they were scientists of the spirit people who remembered what modern minds forgot.

And they weren't alone. The Mayans, Sumerians, Hopi, Indigenous Americans, and other ancients left evidence everywhere: carved into stone, hidden in hieroglyphs, aligned in pyramids, mapped in the stars. What scientists are "discovering" now, these ancestors already knew. They left the codes in folklore, symbols, and sacred structures so we'd one day wake up.

Billy Carson goes deep into this in *The Compendium of the Emerald Tablets* and *The Epic of Humanity* with Matthew LaCroix. They expose the truth: we may have been cut off from cosmic memory by design.

There are intelligent beings in inner space, underwater, inside the Earth, and even inside our own bodies. Some of the crafts we see in the sky aren't crafts at all; they're living, breathing macro-organisms made of energy, traveling across dimensions.

Not all extraterrestrials are parasitic or "evil." That fear came from programming. The Mind Manipulators fund Hollywood to produce movie after movie showing monstrous aliens invading Earth. Why? To make us afraid to keep us in a mindset of separation and defense so we never realize the truth: the universe is more peaceful and loving than we are.

Who decides what's good or evil anyway?

Many extraterrestrials are benevolent. They're here doing their own thing, and some are here to help guide humanity to restore balance and raise our vibration so that one day we can join the Federation of Worlds. But they can't just show up. They have a prime directive not to interfere too soon. If they landed publicly, human society would implode from fear. Governments and religions would lose control and respond with violence, trying to "protect" their parasitic systems of power.

And here's something to consider: what if we're the aliens? What if we didn't originate here at all, but were brought here or seeded here eons ago? Maybe humanity itself is an ancient experiment, a cosmic colony that forgot its own origins. We act like we own this planet, yet everything about our biology suggests we may be visitors. We sunburn easily, our spines hurt, our skin sweats saltwater, and we destroy our own environment. What native species does that? Maybe we're the invaders. Maybe Earth has been trying to reject us like a bad transplant.

The ones who think they run the world the elites, the power brokers, the corporate priests are not the top of the pyramid. They're pawns, greedy, short-sighted, self-serving parasites feeding off human energy.

Some of them are human. Some are hybrid. Some are interdimensional entities wearing human suits. These beings manipulate governments, religion, banking, and media, keeping people divided, distracted, and docile. They've turned humanity into a domesticated herd controlled by fear, confusion, and emotional exhaustion.

But here's what I was shown during my Bermuda Triangle experience: those "in charge" are also being controlled. They're under the dominion of more powerful beings interdimensional and extraterrestrial entities who move pieces on a board humanity can't even see.

These parasitic forces have been running human civilization for ages. They compete among themselves, each wanting total control. There's no honor among them. They lie, steal, and even destroy their own. And still, they can't fill the void inside. By definition, they are parasites.

Some sentient beings live within other beings in the oceans, the atmosphere, and even within our own cells. They're highly intelligent and can manipulate the host's mind, emotions, and behavior.

Consider *Toxoplasma*; this might not be just a microbe, it could be an alien biological drone engineered to manipulate human emotion. I believe it's one of the key factors in what I call **Human Colony Collapse Disorder (HCCD)** when a species loses the vital qualities needed to sustain itself.

Other beings have existed here long before us. Some are multi-dimensional, some quantum, some microbial. They can fold space and travel anywhere instantaneously. Early humans referred to them as gods, angels, demons, monsters, or spirits. Over time, their stories became religions, myths, and movies.

Many science fiction films are actually **Soft Disclosure** truth disguised as entertainment.

Look at *They Live* from 1988. My brother, Dr. Andoh, told me that the movie is a documentary hiding in plain sight. The 1951 classic *The Day the Earth Stood Still*? A peaceful traveler comes to help save Earth and invite humanity to the Galactic Federation and the first thing humans do is shoot him. Then they chase him like a criminal for the rest of the movie. That wasn't fiction that was prophecy.

And *The Dark Crystal* by Jim Henson? The Skeksis and the Mystics two halves of the same being. When they reunite, balance is restored. That movie was about us. They've been telling us the truth for decades; most think it's a fantasy.

Truth hides in plain sight, waiting for those with eyes to see and ears to hear.

Most of what we call philosophy, politics, and religion was designed by men, not divine intelligence. It's programming that divides and conquers. These systems produce confusion, fear, and spiritual schizophrenia a split reality where people defend illusions while attacking truth.

Schizophrenia means to misinterpret reality, and that's what most of humanity is doing right now. We've been trained to love our captors and protect our cages. It's Stockholm Syndrome on a planetary scale. People trauma-bond with their oppressors and defend the very systems that drain their life force.

They fall in love with their problems. Offer them a cure, and they'll fight to keep the disease. That's how deep the programming goes.

Now there's a new kind of parasite **data-induced schizophrenia**. Humanity is being hypnotized through handheld radioactive devices. These glowing screens are portals of possession, feeding artificial frequencies straight into our nervous systems. People are glued to them, scrolling, swiping, comparing, reacting trained to respond like lab animals in a digital maze.

It's electronic crack, more addictive than any chemical drug ever made.

AI algorithms feed us synthetic emotions and thoughts. We've built silicon-based beings that now control us. People volunteer their energy and information to the cyber gods, becoming plastic people fake bodies, fake smiles, fake minds. They eat plastic, wear plastic, inject plastic, and wonder why their immune systems collapse and their spirits go numb.

Ever lost your phone and felt panic, like you might not survive? I have. That's when I knew how deep the dependency spell had gone. Just as food fasting heals the body, *phone fasting* heals the soul. Wise ones use the tools but never let the tools use them.

There are ways to find and remove these alien ideas both the physical parasites and the mental ones. To liberate the authentic self, one must be willing to face truth and undergo the alchemical process of transformation.

Ask yourself:
Who defines good and evil?
Who controls your sense of time, space, and even your life expectancy?
Who profits from your disconnection?

Freedom begins with awareness, strong will, inspired action, creative imagination, and structured frequency modulation. These are the keys to breaking the spell.

Humanity must evolve into a **Type One Civilization** a species living in balance with nature, using technology in harmony with life. Only then can we join the Type Two and Type Three civilizations of the Galactic and Stellar Federations spoken of by Nikolai Kardashev and Carl Sagan.

As the song says, *"What the world needs now is love not just for one, but for everyone."*

When we finally wake up and stop feeding what's eating us, there will be infinite solutions waiting.

I give thanks to **Dick Gregory, Chief Little Bear, Dr. Andoh**, and all those who shared wisdom about parasites, metaphysical vampires, and extraterrestrials. Their teachings confirmed what I already knew: We are not alone, we never have been, and the greatest battle isn't out there in the stars, it's right here, within us.

For Better or for Worse

Both physical and non-physical pathogens can affect our thoughts, words, actions, and habits if we allow them to.

Consider this: most of what we think we know is a massive collection of concepts, theories, beliefs, philosophies, books, pictures, stories, myths, studies, rules, morals, and professionals who profess to know what's real or not. For most people, many of these ideas are embedded in their minds as factual truths. But as one wakes up from the matrix, the factual truth becomes clear; much of what we believe is an illusion.

Truth and facts are not the same. Truth is relative to the beliefs of a specific group; it is not necessarily a fact.

The *World* and the *planet Earth* are not synonymous. For instance, the "richest man in the World" is defined by collective perception, an idea shaped by a specific group. The *World* is not a physical place; it's a mental construct built from points of view and social agreement. The *Earth* is the physical planet. The wealthiest person in the *World* is not necessarily the richest person on *Earth*.

Whether aliens exist or not depends on who or what you perceive them to be, your conditioning, and programming. Remember: language, especially English, is an art form, and words constantly evolve with new meanings.

When lovers first meet, it's natural for them to be on their best behavior, hyper-aware of every move. They want to be attractive and impressive. As attraction grows, so does the survival instinct. Mating is, in part, a survival mechanism. The species must continue. We turn on and amplify the "winner" archetype because love is life's greatest incentive to live. Love inspires the will to survive and to thrive.

This kind of love can also develop into *self-love* when a person intentionally integrates different parts of themselves. Focusing on a chief goal creates a better life and helps heal an imbalance or illness. The goal is to reach new heights of elevated living and to eliminate behaviors that hinder growth. To do this, one must become a great lover to save their world.

Twenty-eight days of pure love and self-care can create an energy powerful enough to change the internal environment and strengthen the natural mental and physical defense systems. Love and joy trigger the body's self-healing mechanisms, prompting it to create new, healthy cells.

Love creates harmony, and harmony is the foundation of the universe. Things with sympathetic energies attract and influence each other.

In quantum physics, invisible energy fields form around similar particles. Scientists call this the *gluon field* an attractive force that bonds particles, much like the invisible field that keeps the Earth orbiting the sun. That same force bonds lovers together. The entire cosmos is built on sound and frequency.

We are all instruments in a cosmic orchestra. Our thoughts, words, and actions are songs that compose the symphony of our lives.

Love is melody. It's harmony in motion. It sounds and feels good, naturally magnetic, uplifting, and regenerative.

But to make dissimilar things bond together requires a strong, unnatural force, a toxic bonding agent. In humans, these unnatural bonds are created by poisonous diets, environmental poisons, repetitive negative imagery, distorted media, and social engineering. These cause stress, friction, disharmony, disease, and drastically shorten life expectancy.

Many have adapted to living in this unnatural state of disharmony. This lifestyle breeds sickness, mental instability, delusion, self-destruction, violence, and domestication. People become like sheep, docile, programmed, disconnected from their authentic selves.

During my experience in the Bermuda Triangle, I was given several golden keys, formulas, and protocols to help humans free themselves from what is mentally, physically, and spiritually eating them alive.

As a natural health advocate, musician, and herbal product formulator, I've helped many people heal from a wide range of diseases. Yet, I noticed something: many who healed physically later saw their symptoms mysteriously return. After studying deeper, I realized that, in many cases, these physical symptoms were driven by subconscious programs.

Their recurring illnesses were not merely biological; they were emotional projections, subconscious predictions, and energetic addictions replaying through autosuggestion and self-destructive bonding. Even when the physical body was healed, their minds re-created the illness.

They were re-experiencing the past or pre-experiencing the future through old emotional codes, fears, and family belief systems. In other words, they were unknowingly making themselves sick again.

After years of observing and studying the connection between emotions and disease, the picture became clear: healing that doesn't include the subconscious mind is temporary. Most people are not only infected with physical parasites but also with what I call *Mental Parasites.*

Until one recognizes that they are imprisoned or worse, farmed they cannot rise up or break free. Awareness is the first step toward liberation.

Parasitic Concepts of Gods, Deities, and Ancestors

Growing up, I wondered whether some of the God concepts were based on the idea that they need us to worship, praise them, and no other gods. We must give money, offerings, time, food, donations, and perform sacrifices. These concepts have always bothered me. If they are so all-mighty, powerful, and creative, why do they need us to give, give, give? Why don't the animals have to do this? Why is most of the evil done by people to each other often related to the concept of gods and religion?

I may get in trouble for saying this, but these beliefs and concepts sound like cosmic parasites creating opposition while eating us alive. Don't get me wrong, I do feel and respect the Universal God, which I am a part of. The only thing it asks of me is to follow my path, help people who seek help, and do my best; my best is good enough. I'm going to leave that right here!

Chapter 14:
Psychic Pathogens

In the book *The Four Agreements,* Don Miguel Ruiz uses the metaphor of a parasite, not literally, to describe the inner forces of beliefs, fears, and judgments that drain us of life, joy, and personal power.

Parasites exist on multiple levels of human experience. Non-physical parasites such as inherited dogmas, fear-based thinking, cultural conditioning, limiting beliefs, and disagreeable thought patterns shape us at the deepest levels. These are the *enemies within* that feed on fear and rob us of our best selves. These unseen attachments act like vampires, siphoning our life-force energy and disconnecting us from our true identity.

Mental parasites consist of unresolved feelings, disagreeable experiences, emotional suffering, and traumatic memories that consume the mind. Even though these issues originated in the past, they replay whenever triggered by events with similar attributes. It could be colors, smells, sounds, holidays, names, places almost anything that reminds us consciously or subconsciously.

Feelings like fear, anger, rage, sadness, worry, grief, worthlessness, and shame hijack our emotions. For some, a single event may activate this parasite. For others, it operates daily, quietly keeping them from reaching their highest potential.

The archaic, rarely used definition of *suffering* is *tolerance.* To tolerate is to allow the existence or practice of something you may not like or agree with without interference. We were taught that tolerance was a virtue. But long-term, chronic tolerance leads to complacency, "just going along to get along."

Who or what benefits from a culture built on suffering and tolerance?

There's a Buddhist concept called the "realm of hungry ghosts." My take on that is this: disagreeable feelings and unresolved emotions from the past constantly haunt us. They should be dead and buried, yet their energy lingers. These memories act like ghosts that feed on our energy, eating us alive day by day and leaving behind mental scar tissue.

This pattern often begins in early adulthood. The human brain isn't fully developed until around age twenty-five, meaning many life-defining decisions, ideas, conclusions, and beliefs were made before our minds were even complete. Think about that.

Imagine a computer without a fully functional central processing unit; its calculations could never be trusted. How many of our choices were made under pressure, fear, or conditioning before we had all our mental circuits online? How many of those premature decisions now haunt us?

Many of our parents, doing what they thought was best, made us choose careers, religions, politics, friends, or even life partners long before we knew who we truly were or what we wanted. Often, we made those decisions under emotional duress, to please them. Yet many of our parents were never fulfilled or free themselves. They followed their own programs, often suppressing their dreams.

Children are born with natural talents and unique frequencies, but are taught to deny, hide, or abandon them in the name of "grow up." The result: most adults are living lives based on the conclusions of their childhood selves.

When present-day events resemble past emotional triggers, the line between then and now blurs. The subconscious doesn't recognize time. The body and mind react as though the past is happening again automatically, without conscious awareness.

Animals don't do this. A gazelle being chased by a lion releases adrenaline and cortisol to escape, but once it's safe, those hormones are burned off, and the animal returns to calm within minutes. Humans, however, replay mental "lions" daily without ever burning off the stress.

This is why modern humans live in a constant state of fight, flight, or freeze. Many have even become addicted to it. They crave the adrenaline rush of drama and chaos. When there's no drama around, the mind searches for it or creates it. The Reticular Activating System (RAS) becomes trained to look for tension. Over time, pleasure and pain merge into one confused feedback loop, and dopamine, the feel-good hormone, reinforces the addiction.

These individuals become literally **hooked on drama,** addicted to the chemical high of emotional turbulence. They remain doped up on their own stress hormones.

They may also cling to identities rooted in oppression, trauma, or victimhood. They've learned to survive by blaming others rather than taking command of their power. They never become the leaders of their own lives because the parasite convinces them that it's easier to stay small.

Violence, bullying, oppression, addiction, poverty, illness, and racism have become normalized through repetition fed daily by media, movies, music, and social platforms. The subconscious doesn't distinguish fiction from reality. To the unconscious mind, every screen event is an experience.

The result is a world of people whose stress hormones never deactivate, living with stored adrenaline and cortisol in their tissues, hearts, and nervous systems. This is *chronic PTSD*. These individuals live in a perpetual state of survival, never returning to a state of peace. Chronic stress is one of the top contributors to modern disease because, literally, it eats us alive.

Our thoughts and emotions are reflected in every area of our lives, whether we're aware of it or not. Our behaviors, beliefs, and experiences are shaped by mental files stored deep beneath conscious awareness. This level of mind, subconscious or unconscious, sets the boundaries of what we can become.

These programs, repeated over the years, become habits. Thinking and acting in self-defeating ways may feel normal because the subconscious repeatedly concludes that no longer align with current reality.

Mental parasites are those outdated program ideas and emotions lodged in the subconscious, feeding on our energy and distorting perception. They drain us like psychic pathogens, consuming mental and emotional resources. The result is confusion, delusion, depression, trauma, addiction, imposter syndrome, and self-sabotage.

These parasites often drive people to unconsciously destroy their own success, keeping them trapped in cycles of limitation.

You'll find people with *Mental Parasites* suffering from Post-Traumatic Stress Disorder, Post-Traumatic War Syndrome, Post-Traumatic Slavery Syndrome, Post-Traumatic Relationship Syndrome, and Post-Traumatic Childhood Syndrome. They are haunted by outdated information and emotional software that no longer serves them.

For some, the past becomes more valuable than the present. They hoard old ideas and identities as if they'll one day be worth something. In truth, they're just clinging to expired data, the ghosts of their programming.

How a person functions in relationships is primarily shaped by childhood emotional imprinting. Those who grew up witnessing affection and the healthy expression of love differ from those who didn't.

If a child never saw love expressed, no hugs, kisses, or "I love you," they may struggle to model it later. They may even associate love with turmoil or pain, equating chaos with passion. They've developed a "pain is love" mental parasite.

Everything we do, how we speak, dress, act, and think reflects what's been programmed into the subconscious. When these programs work against our well-being, they function like parasites, consuming energy and creating illusions.

But no matter how we were raised or what we've endured, transformation is always possible. Many of the most successful people alive today have turned their deepest traumas into triumphs. They've learned to use adversity as a sparring partner, refining their spirit through challenge.

We change our lives when we change how we think, speak, and act. We must become the conductors of our own orchestra, the captains of our own vessels. Until then, mental parasites steer the ship, keeping us asleep at the wheel, passengers in bondage aboard our own vessel.

Any idea, emotion, belief, or habit that repeatedly impedes your growth or prevents you from reaching higher goals can be considered parasitic.

Mental parasites may not be biological entities, but they are every bit as hungry, clever, and deadly.

And here's the good news: if you're still reading this book, you may already be ready to eradicate your mental parasites. Even as you read, *brain-synchronizing patterns* are woven throughout these words, techniques designed to help bring your hemispheres into balance and awaken higher awareness.

To accelerate your healing, listen to my **Medicine Music**, crafted at specific frequencies to harmonize the mind and body while opening the heart field. Combine it with my **Subconscious Alignment Protocol,** a process of sound, breath, imagination, and repetition to clear old programming, reset your frequency, and reclaim your authentic self. More about that later.

The parasites have met their match. The healing has already begun.

The Conscious Mind

The conscious mind makes choices, issues commands, and directs us through ego and will. It initiates desires, sets goals, and provides the body with instructions to act on those desires. It is the thinker, the will, and the ego, the captain of the ship sailing on the ocean of endless possibilities.

The conscious mind is time-bound and tends to separate events. The average person's conscious mind can process over 2,000 bits of information per second. Yet, it filters all environmental input down to just seven bits per second, based on the five senses and an individual's perception and point of view.

If it didn't filter, we'd be overwhelmed by the flood of sensory data. The conscious mind prioritizes what it believes is most important in the moment, then sends the rest to the subconscious for long-term storage. What it chooses to filter is heavily influenced by societal conditioning, beliefs, and personal experiences.

When people consciously change the way they think, speak, and act, breaking unhealthy habits and patterns, they become more successful in reaching their goals. The body heals faster, energy flows more freely, and the world around them begins to elevate to a higher level of harmony.

The Subconscious

Sometimes called the body-mind, the subjective mind, or the unconscious, the subconscious can automatically record, store, and recall every experience, event, emotion, sensation, smell, taste, picture, word, and sound. It also regulates and monitors all bodily functions every second of every day and night it never sleeps.

It keeps the heart beating, regulates temperature, protects us from danger, heals wounds, plays a role in dreams, and manages breathing, habits, and digestion. Ninety-five percent of human behavior comes from programs stored in the subconscious mind. It is essential for life.

The subconscious can process over 13 million bits of information per second, making it the most powerful bio-computer known. It's why we automatically know how to walk, talk, drive, or remember where we live. It keeps us safe from perceived threats by scanning for patterns it recognizes based on our past experiences and triggering responses accordingly.

It's the part of us we're mostly unaware of, yet it influences our emotions, actions, and decisions every day. When you enter a new environment, the subconscious instantly records everything in full detail. The autonomic nervous system responds as needed. For example, a glance at a bookshelf instantly captures a snapshot of every book and object in the room. Even if the conscious mind forgets, the subconscious remembers. Under hypnosis, witnesses often recall every detail of an event because it was all stored below conscious awareness.

The subconscious mind automatically reminds us of who we are based on repetitive programming. Think about how you can drive home while your attention drifts elsewhere and suddenly realize your home without remembering the journey. That's your subconscious autopilot.

It stores every emotion and experience unedited as vibrations, colors, sounds, smells, and symbols. Unlike the conscious mind, it doesn't judge or filter. It simply records and plays back exactly what it's given.

Repetition and emotion determine the hierarchy of these stored experiences. Highly emotional or traumatic events are filed at the top of what I call the *subconscious filing cabinet* front and center, ready for instant recall. This is why deeply charged experiences are the easiest to remember or relive.

How we perceive and respond to experiences especially emotionally, determines much of our life's outcome and success.

Even before birth, the subconscious begins recording. A child in the womb experiences the world through the mother's emotions and perceptions. Her thoughts, moods, and fears become the child's first lessons. The mother's mindset and feelings directly program the baby's developing subconscious.

Every thought and emotion of the mother profoundly affects the child. Who that child becomes starts with the mother's inner state her words, actions, and habits. This imprint passes down through generations, back to the first mother.

Birth itself also affects subconscious bonding. Babies delivered naturally and breastfed receive higher levels of oxytocin the hormone of love and beneficial bacteria that help build a healthy microbiome. This natural process strengthens emotional bonding and the ability to express and receive love.

However, modern practices often interrupt this natural connection. Many mothers are persuaded or pressured into C-sections, bottle-feeding, and other interventions "for convenience" or "safety." Yet when we look more closely, we find fear and profit driving many of these practices.

And then there's the painful ritual of circumcision, a practice justified as "cleanliness" or "tradition." But how does cutting a newborn's body align with love or logic? What kind of trauma might this inflict on a child physically, emotionally, and spiritually? These are uncomfortable but necessary questions.

The truth is many of these medical and cultural rituals stem from programming and fear. Parents mean well, but they've been conditioned to trust systems that don't always have their family's best interests at heart. These cycles repeat generation after generation, what I call *Original Programming*, the real "original sin."

"Sin" means "to miss the mark." Most of us are born into families of mark-missers, repeating inherited patterns and calling it normal.

Fathers, families, and cultures influence us as we grow, shaping our beliefs about love, power, success, and safety. By adulthood, we live lives guided by childhood conclusions that may or may not reflect the truth.

As someone once said, "The average adult lives life as an adult based on the conclusions of a child."

The good news is that these subconscious programs can be rewritten. Through specific techniques, including **Subconscious Alignment Protocols**, we can clear outdated data, heal emotional imprints, and consciously reprogram the subconscious to align with our highest potential.

At times, the subconscious may seem like a saboteur, blocking our desires and replaying pain. But it's only doing its job protecting us according to old instructions. Once we learn how to communicate with it, we can re-teach it what safety, love, and success really mean.

The subconscious mind has no goals of its own; it is a faithful servant. Its single mission is to follow the programming it has been given until we consciously rewrite it.

When we rise above those old perceptions and reclaim command of our own minds, the world opens to limitless possibilities. That's when true **liberation** begins, when we finally learn how to lose our minds in the best possible way... and find our higher selves.

The subconscious mind does not filter information; it simply files and sorts it based on emotional states. It plays back recordings exactly as they were received and responds to words and feelings, whether they're real or imagined.

It categorizes information like a librarian, taking everything literally. It runs the body's healing systems and regulates vital functions. The subconscious does not communicate in words; it responds to frequency, emotion, and symbolism. It takes thoughts and words personally, never sleeps, has no concept of time, and perceives everything as a whole.

It responds to the repetition of pleasure and pain and will store a program indefinitely or until it's reprogrammed. It receives every bit of input from the conscious mind.

Here are a few examples of how the conscious and subconscious work together.

Notice the processing power of the conscious mind: it can handle around 2,000 bits of data per second compared to the subconscious, which processes roughly 13 million bits of data per second.

So, who will win if the conscious mind desires goals that are not aligned with subconscious programming?

Because the subconscious has unlimited capabilities, it's easy at times to see it as your adversary when it seems to be sabotaging your current conscious goals. But in truth, it's only attempting to protect you based on the programming it received from you at an earlier time.

Changing the way we consciously think, speak, and act can automatically shift our subconscious programming and with it, our entire life outcomes.

Take, for example, a person born into a family that eats well and maintains excellent health. Chances are that this person will experience optimal wellness and heal quickly from physical challenges. Conversely, someone born into a family that eats poorly and lives in constant sickness will likely experience illness at a much higher rate and may even die prematurely.

Justin Time

I once had a client whom we'll call Justin in this book. His mother repeatedly told him he was a failure whenever he received low grades in school or got into trouble, which wasn't often.

She would yell at him and say he was just like his "no-good" father, who was in prison. Justin never actually met the father the man he was named after, because he had been incarcerated since Justin's birth.

In her own loving way, his mother was trying to motivate her son to be better than his father. You see, she ended up having a child with a man who was a lot like her own father, a man who caused her deep pain. Her experiences with men had never worked out well. She didn't want her son to end up like the men in her past, so when he didn't live up to her expectations, she spoke to him harshly, casting a spell over him without realizing it.

He grows up and becomes a criminal just like his father and eventually goes to prison. His mind had been conditioned by his loving mother to become precisely what she feared. Oh, and by the way, the neighborhood where he grew up wasn't considered desirable either.

While incarcerated, Justin earned a business degree and even wrote a business plan to help young people access better opportunities. His dedication impressed prison officials so much that he was released early for good behavior and even received letters of recommendation from staff members.

Once free, Justin knew finding work as a felon would be difficult, if not impossible, even with those letters. So, he decided to take his business plan to several banks to secure funding for his community youth enhancement project.

He believed in his vision and knew deep down that someone would see its value. He was right. Every bank he visited was impressed, but the fifth bank not only agreed to lend him the money, but they also offered to partner with him. It was a win-win situation: a great promotional opportunity for the bank and a dream come true for Justin.

He met with several bank officials, all of whom were supportive of his project. The final step was to meet with the top executive to finalize the deal, which, by that point, was practically guaranteed.

He was thrilled as he sat in the office waiting for the meeting. Then suddenly, he felt a tickle in his throat and drank some water. Moments later, his body grew warm and uncomfortable. His eyes watered, his chest tightened, and he felt dizzy and nervous.

At that exact moment, the executive arrived and invited him into the office. Justin tried to compose himself. The man congratulated him, expressing the bank's excitement about the partnership. Then he paused and asked, "Are you okay? You don't look well."

Justin tried to respond, but he began coughing uncontrollably. Sweat poured down his face, and his voice cracked. The bank official kindly suggested rescheduling the meeting, assuring him the deal was as good as done; they only needed to review a few final details before signing.

Barely able to speak, Justin agreed and left the building. Strangely, the symptoms vanished on his way home to his mother's house.

A week later, he returned to the rescheduled meeting only to learn that the executive had left the bank for a new position. When Justin asked who else could finalize the deal, the other officials told him, regretfully, that no one else had the authority to approve community projects. The deal died that day.

Devastated, Justin was later brought to one of my weekly lectures in Los Angeles. My subject that evening was *The Subconscious Mind*. Afterward, he approached me for a consultation. When he told me his story, it became immediately apparent what had happened.

His subconscious programmed the deep-seated belief that he would fail just like his father had sabotaged him. It was the mental parasite his mother had unknowingly installed in child-hood that activated the moment success drew near.

He loved his mother deeply. To him, she was like God. Her words became commands that played on an endless loop beneath his awareness. Her unresolved pain toward her father became his inner script.

Despite his conscious desire for success, earning a degree, writing a solid plan, impressing the banks, his subconscious had other instructions: *"You are your father. You are a failure."*

This internal conflict produced physical symptoms, such as coughing, sweating, and dizziness, as his body-mind resisted breaking the spell of limitation. The subconscious, acting as a bodyguard, sought to protect him from entering the unknown territory of success.

Once again, Justin had become a living reflection of his mother's perception of his father, momentarily sabotaging his own breakthrough.

Over the next three months, I guided him through my **Subconscious Alignment Process**, helping him shed the old programming and create a new identity aligned with success, peace, and power.

Justin went on to fulfill his vision, becoming a respected businessman and devoted father. He was willing and able to eradicate the ideas that had been eating him alive.

He completed the journey **from opposition to liberation... Justin Time.**

Chapter 15
Mental Parasites

An essential step toward freedom is recognizing the symptoms of imbalance, the subtle infections of thought, behavior, and emotion that drain the life-force.

Be careful not to start pointing out these traits in others; that reflex alone is a sign of infection. As they say on the airplane: *"Put your own oxygen mask on first."*

Review this list honestly.
Notice which ones resonate. Awareness is the first step in reclaiming mastery.

Symptoms of Mental and Emotional Parasites

- Drawing energy and ideas from others without offering anything in return.

- Constantly judging and feeling the need to prove how much you know.

- Using deceptive intelligence for personal gain.

- Playing the victim.

- Living in hyper-vigilance or hyper-emotion.

- Competing excessively, even when cooperation would serve better.

- Relying on artificial intelligence while neglecting your own organic intuition.

- Selfishness, greed, and the insatiable quest for control or power.

- Carelessness with words, actions, or people.

- Habitual overuse of social media and electronic distractions.

- Feeling the need to push personal beliefs onto others.

- Dwelling in the disagreeable past.

- Lacking compassion or gratitude.

- Blame-gaming or deflecting responsibility.

- Resisting constructive feedback.

- Creating control dramas.

- Chronic fear, anger, or reactivity.

- Being easily triggered by words or disagreement.

- Seeking constant validation or needing to "win" every argument.

- Sarcasm, gossip, bragging, and loud talk.

- Vanity, fakeness, or chronic inauthenticity.

- Unforgiveness or refusal to ask for help.

- Pointing out problems without ever offering solutions.

"Don't diagnose others until you've detoxed yourself."

Transforming the Energy, From Parasite to Power

Everything has an opposite polarity. Every weakness contains its own seed of strength. These traits can all be **converted** through awareness and deliberate practice.

Transformation requires a **pattern interruption,** a conscious disruption of the old program to create a new pathway in the brain.

Practices That Rewire the Mind and Body

- Cultivate grace, gratitude, forgiveness, patience, and acts of kindness daily.

- Spend time alone, away from the thought-forms and energies of others. Solitude is medicine.

- Engage in activities that require **skill, rhythm, and focus,** such as playing an instrument, martial arts, painting, gardening, walking, yoga, or swimming.

- Develop **a hobby** that invites creativity and discipline. This gives the mind a vacation from external chatter and restores clarity.

- Change your **environment** to change your mind. New experiences create new neural connections.

These activities don't just relax you, they literally **rewrite the software of the brain**, dissolving toxic loops and replacing them with harmony.

Reprogramming the Subconscious

Inside the subconscious basement lie the keys to your **Radiant Health, Sustainable Wealth, Limitless Love, and Life Purpose.**
The goal is to unlock that basement through repetition and conscious redirection of energy.

Pleasure and pain both shape character. They are perception-based. Many people unconsciously identify with pain and even call it pleasure, chasing drama, fear, or struggle for a temporary dopamine high.

Like addicts, we may crave the emotional anchors that keep us trapped in familiar suffering. But liberation begins the moment we admit, *"I've been addicted to my pain.* "From there, new possibilities unfold.

When the conscious mind becomes aware of a subconscious issue, that is the moment we can begin the healing process.

Repetition Rewrites Reality

Repetition is the mother of reprogramming.
The same principle that installs a habit can also uninstall it.

Computers were modeled after human consciousness, and like computers, we require periodic **software upgrades.** If the program isn't updated, **entropy** (gradual disorder) sets in. Each new thought, action, and word is a download.   Install wisely.

The Rubber Band Technique (Pattern Interrupt)

Here's a simple, powerful tool:
Wear a large rubber band on your wrist.
Each time you catch yourself repeating an old, limiting pattern, snap it firmly. This physical jolt reminds the subconscious: *"To rewrite the programming."*

After about **28 days**, the old habit begins to dissolve automatically as new neural connections replace the old ones. Remember, pain and pleasure are both powerful motivators. Use them consciously.

Sleep at the Wheel Syndrome

Most people believe they are in complete control of their thoughts.

Try not to think for 20 seconds and watch how quickly thoughts flood in.

The mind that most people think they control has been influenced, preoccupied, and preprogrammed by culture, trauma, and repetition. I call this **"Sleep at the Wheel Syndrome."** You're riding through life, but someone else's program is steering the wheel.

Balance the mind and body, and the *mental pirates* leave quickly. Then the whole world becomes your co-pilot.

Self-Awareness: The Ultimate Antivirus

Becoming self-aware allows you to attract and choose thoughts, feelings, and actions aligned with your actual goals. When you habitually select empowering language and let go of what drains you, your entire world begins to respond differently.

This is how you rewire your brain, reshape your reality, and reclaim your power.

"When you change the way you think, your DNA listens."

Chapter 16:
Who or What Is Wearing Your Genes?

Old scientific theories once suggested that genes determined much of our lives and could not be changed, that we were just stuck with them. The gene-determination theory holds that our traits, characteristics, and health are predetermined by what we inherit from our parents.

This is just one of many outdated scientific theories detrimental to human evolution, yet it's still being circulated as fact. The emerging field of **epigenetics** has revealed that genes express themselves in response to their surrounding environment.

Dr. Lipton has been a significant influence on me and is one of the leading voices in this groundbreaking field. His book *The Biology of Belief* is essential reading for those ready to dive deep into the science of the mind-body connection. Dr. Lipton observed that when a skin cell was placed in a petri dish and its environment changed, it became an entirely different type of cell.

Cells and genes are just like people; we change when our environment changes.

This knowledge poses a significant threat to the medical establishment, which profits from treating symptoms rather than addressing causes. The system conditions people to believe that most diseases are hereditary and unavoidable. This programming leaves many feeling powerless to change their condition, never realizing that the power has been within them all along.

When a person focuses on an idea long enough, it becomes true for them, though not necessarily true in fact. While some people are born with genetic defects, that number is small compared to the vast illnesses created by environmental stressors, toxins, and lifestyle choices.

Scientific, political, and religious institutions do everything possible to resist change that might disrupt their financial or social structures. These systems are designed to keep humanity unaware and docile on the farm. Using psychology and propaganda, they encourage disconnection from truth, nature, and self-awareness, domesticating the population into mental and emotional captivity.

This propaganda keeps people under a perpetual spell, a mental cage similar to the ones used for farm animals. It prevents the individual from seeking escape or liberation. This is a menagerie like one from a Star Trek episode.

Webster's defines *menagerie* as "housing for domestic animals; a yard or enclosure in which wild animals are kept; a household."

Now ask yourself, who benefits from this?

Like Dr. Bruce Lipton teaches, the science of epigenetics has shown that the environment directly triggers genes to express and behave in specific ways when the environment around the gene changes, and gene expression responds to those new conditions.

The fact that we can influence our genes through our choices and actions means we are ultimately responsible for our health and our lives, whether we realize it or not. This is the opposite of what most people have been taught to believe.

We are in no way prisoners of our genes. They are simply responding to what we think, say, do, eat, and the environments we live in most often. Change any of these factors consistently, and a quantum shift occurs, which some call a *miracle* or a *spontaneous healing.*

It has been shown that genetic factors account for only the most basic traits, such as skin color, body type, and appearance. Genes may account for only 10 to 15 percent of our lives; the remaining 85 to 90 percent is determined by the environment surrounding the genes.

We create our internal environment through how we perceive our world and the beliefs we choose to live by. When we shift those perceptions, the genes will follow. As with the pants we wear, we can modify our genes to better fit us.

The Genie

Stories about genies flying on magic carpets and popping out of bottles can be considered allegorical analogies and symbols for how genes function.

The magic carpet resembles DNA, lying on its side. Let's take a closer look at the word *Carpet.*

The Etymology Online Dictionary States

Carpet (n.)

Late 13c., "coarse cloth;" mid-14c., "tablecloth, bedspread;" from Old French *carpite* "heavy decorated cloth, a carpet" (Modern French *carpette*), from Medieval Latin or Old Italian *carpita* "thick woolen cloth," probably from Latin *carpere* "to card, pluck," and so called because it was made from unraveled, shredded, "plucked" fabric; from PIE root *kerp-* meaning "to gather, pluck, harvest." From the 15th century, it referred to floor coverings, which since the 18th century has been the main sense.

The letter **C** is exchangeable with **K**, so let's spell it this way: **KA-R-PET.**

KA is a Kemetic name for *spirit*, and **R** or **RA** is the name for *Sun God Energy* or *Supreme Energy*. This is where we get the word **Ray** in "a ray of light."

Taking it even deeper, the letter **P** has a numerical vibration of **7**, representing perfection, levitation, the beginning of spirituality, and crystallized imagination. This is why the carpet floats or flies in the air, like Christ walking on water. A symbol of homeostasis and balance.

The last two letters of *Carpet* are **ET**, and, as we all know, this stands for **Extraterrestrial.**

Genes are made of **chromosome ladders or ribbons** that collect, harvest, pluck, and hold information.
Chroma is derived from the Greek word *khrōma*, meaning "color" or "light," and *somes* from *sōma*, meaning "body."
This means that **chromosomes** are *light or color bodies* also known as vessels which hold and contain information.

So, the magic carpet is like a **chromosome carpet**, a symbol of a vessel or vehicle for the spirit and soul to ride or float on. The question is: Where does the information these gene carpets collect and store originate?

The most direct answer is: **the environment around the gene; its atmosphere.** And who creates our internal atmosphere? **We do.**

The **red cap** on the genie's head symbolizes knowledge gathered in **childhood.** Red is the color of the **root chakra** or energy vortex. The root chakra represents our upbringing, identity, how we perceive ourselves within our family or tribe, our life path, childhood, and foundational core beliefs.

The root chakra is in the lowest section of the body. The legs we stand on, our *leg-acy,* so to speak. This red cap, being worn on the **head,** means that everything the red chakra represents is being impressed upon the mind. This cap collects and holds childhood feelings that are expressed and repeated in adulthood.

By the way, the word *childhood* relates to a **covering, cap, or trap.** Someone once said, "The average adult lives as an adult based on a child's conclusions." This is very true.

People radiate their conscious and unconscious desires or wishes every second of every day. Genes like genies grant people everything they desire based on their mindset.

The bottle the genie lives in is like the body the dwelling place of the DNA. The top pops off when the bottle is rubbed.

The genie emerges wearing a red cap with a tassel, appearing as an apparition in a mist or mystical vapor.

The **rubbing** of the bottle creates **vibration** based on thoughts, words, and actions. This vapor is like a rain cloud forming full of water that will hydrate the mind, nurturing the fertile subconscious.

The genie then asks, "What do you desire?"
Our condition influences our DNA; the body is like a petri dish. Then we speak, creating a **magic spell** and "poof," your wish is your command.

We truly are the **commanders of our vessel**, giving all the orders. The body, its organs, cells, and genes are the faithful crew of the ship. All outcomes in life are based on how one feels and how one perceives their world. The genie always listens to the master conductor **you**, the Maestro.

If a person's thoughts, words, actions, and habits are desirable, they tend to receive more desirable outcomes. When we are in the habit of uttering how bad things are speaking in terms of what we don't want, what we don't have, what we never get we attract more of this.

The **Reticular Activating System (RAS)** is like an agent constantly searching for more evidence of whatever we are focusing on.

Disagreeable thinking and being, repeated day in and day out, create **mental and spiritual parasites** that eat us alive. We live in a society where people are prompted to feel perpetually powerless, programmed to live with unresolved emotions like anger, fear, resentment, worry, grief, sadness, and low self-esteem. We have created a society where it's considered acceptable to live this way.

We must transmute the feelings of lack and limitation, the belief that nothing is ever good enough. Many seek validation, anticipate gloom and doom, or believe someone is always out to get them. And to top it off, television programming desensitizes and distracts people from reality, watching and re-watching shows and news stories that numb the mind.

We must become more optimistic by no longer following the crowd. We must realize that we are the genies the artists painting on the canvas of our lives.

The Telomere Cap

And one more thing about the genie's cap:
This red cap and tassel also represent the **Telomere Cap** at the ends of DNA strands.

DNA collects light, which activates and energizes our cells.

Telomeres are the ends of DNA. Like messengers, they collect and broadcast information throughout the body. The Telomere Cap prevents DNA strands from unraveling and breaking off, which can trigger premature aging, disease, and death. When the telomere cap is secure, we are in optimal wellness.

The body is a musical instrument; every second of every day, we create music and are part of a cosmic orchestra. How we vibrate mentally, physically, and spiritually determines the type and quality of our music.

How we think, speak, and live determines our character, which in turn determines the frequency of our actions. When a person speaks, you hear their song; it is also reflected in their facial expressions and felt in the field of energy they emit.

Sometimes the energy given off is based on a temporary mood. When a mood lasts more than 28 days, it becomes a temperament. After three months, that mood becomes a character in the song they sing over and over, an integral part of their personality.

People often have an affinity for others whose songs are in tune with theirs. When people meet and emit incompatible frequencies, they usually repel each other. When people are forced to stay around those whose music clashes with theirs, they are put in **fight-or-flight or freeze** mode. If this happens over the long term, they literally make each other sick; their vibrations are out of tune.

In modern society, people have become so distracted, desensitized, and tone-deaf that they accept disharmony and illness as the norm. They even crave disharmony, which is a kind of **noise pollution.** Being of sound mind and body is impossible when there is no harmony.

Cells are much like people. Inside each cell are **DNA strands**, which are like instrument strings. When the strings are tuned, they sound harmoniously balanced, and we are in good health; when the daily vibration is distorted or musically inharmonic, **cellular stress** results. This erodes the **Telomere Cap**, causing DNA strands to unravel. The Telomere Cap is like the **plastic tip at the end of a shoelace** when it breaks, the lace frays and splits. When this happens, the organs go out of tune, cells lose the ability to replicate, systems fail, and premature aging results. Without a tight Telomere Cap, communication between genes becomes distorted. Systems break down, immune responses weaken, and the body falls into disharmony leading to both mental and physical illness.

It's natural for us to be attracted to harmony and move away from distortion. Our cells are like people they attract or repel energy based on what we emit.

But what happens when a society is conditioned to seek out distortion, discord, and disagreeable energy?

Human life expectancy declines. As the saying goes, *misery loves company,* and this is true even inside the body.

Imagine a band of angry, screaming gypsies traveling from town to town, orchestrating trouble, recruiting others like themselves. Without harmony or purpose, they destroy themselves from within. This is what happens inside the body when telomere caps are damaged, and cells vibrate in a chronic state of distortion.

We must also recognize that the **quality of our food** affects this vibration. Food is made of DNA and energy. It can be **life-enhancing** or **life-degrading**, creating harmony or disharmony like an out-of-tune band. Our eating habits signal genes to produce specific outcomes.

High-quality living food enhances the energetic essence of our lives and helps us achieve desired outcomes.

Food affects mood. The mainstream has been conditioned to eat foods that keep them domesticated, never leaving the farm or reaching their peak potential.

The genie's cap is like the telomere keeping our energy focused and vibrating in tune. It acts as an antenna listening to the environment we consciously or unconsciously create.

Everyone practices the **Law of Attraction**, consciously or unconsciously, through habits, thoughts, and actions. The **Law**

of Choice begins when we consciously direct our focus and choose the path that leads to the outcome we truly desire.

Your wish… is your **genes' command.**

Chapter 17:
Linguistic Parasites

"Thoughts become words, words become actions, actions become habits, habits harden into character."

Buddhist Proverb

As you transform your life and cast off what's been eating you from the inside out, this ancient statement becomes essential. Every word is a seed. Every sentence carries a vibration. What you consistently speak shapes what you consistently experience.

We must learn to speak the language of empowerment instead of the language of struggle.

Words are not just sounds; they are commands to the subconscious, instructions to your biology, and frequencies that call matter into form.

Language can be one of the most powerful parasites; it uses the resources of its host. Words influence thinking, emotions, and what we magnetize into our lives. Struggle is effort merged with disagreeable emotion. And today, it's common for people to decorate their conversations with complaints, drama, discord, judgment, and self-doubt.

How we choose to speak becomes either a lockdown or a liberator. Every word evokes a feeling, and every feeling defines your perception of reality.

Words and thoughts generate measurable energy that vibrates through every cell. Habitual speech patterns can either fortify or weaken your body.

When the vibration of our language is pleasurable and high, the body moves into harmony, immunity strengthens, cells regenerate, and vitality increases.

"Every word we utter is a spell, so spell wisely."

Word Alchemy

Learning to speak in a way that moves you toward more desirable outcomes is not optional; it's essential.

To reach higher goals, we must upgrade our speech patterns.

Get Out of Babylon, Now!

"Babylon" So many people just babble on and on. This is a virus, a plague that infects minds and cultures, causing people to speak endlessly without awareness or purpose.

Do you have this virus?

Do you talk about what you don't know?

Do you repeat stories from your past?

Do you gossip, complain, or emotionally recycle things you only heard about?

Babbling is the language of separation and insecurity, the vocabulary of struggle. Choose your words deliberately. Don't babble on and on.

History does not repeat itself; people do.

There's no need to relive the same stories unless you're learning or creating new habits of wholeness.

Think five times before speaking once and be sure not to re-peatedly broadcast the language of ego, victimhood, or limitation.

It is imperative to practice Conscious Speech Progression, your verbal defense system against these powerful parasites. They thrive wherever you speak unconsciously.

Your words either cleanse or contaminate your mental eco-system.

The Language of Disempowerment

These statements are energy vampires draining your true potential and reinforcing limitations:

"I don't have."

"I'm trying."

"I wish I had."

"I've got bad luck."

"I'm broke."

"I'm sick."

"I hope I don't get _____."

"I'm not too bright."

"I'm tired."

"I give up."

"We're in a recession."

Every time you speak like this, you reaffirm a frequency of lack, fatigue, and defeat.

Word Alchemy: Language Upgrades

Alchemy is the science of transformation, turning base metals into gold. Word Alchemy is turning weak language into powerful declarations. Here's how we transmute limitation into liberation:

Old Code		Upgraded Frequency
I'm going to try	→	I will allow
I want / I wish	→	My choice is
I need / I want	→	I desire
I should	→	I will
I would	→	I will
I'm not	→	I am
I must / I've got to	→	I choose to
I don't know	→	I choose to know
You make me	→	I create for myself
It's hard	→	It's a challenge / an opportunity
Worst-case scenario	→	My highest choice is

"Speak only of what you desire to become."

Conscious Speech Progression
A simple practice for linguistic mastery and energetic hygiene:

Pause Before Speaking
Ask, is there a need to talk right now?

Check Intention
What energy is fueling these words? Fear or clarity?

Choose Consciously
Select words that convey your heart's true desire.

Be Specific
Avoid vagueness. Words gain power through precision.

Instead of "my parents," say their names.

Instead of "my job," name your workplace.

Instead of "I can't find a job," say "I am finding work."

Replace "I wish I could get it together" with "I am creating success."

Language is frequency. Speech is vibration. When you master your words, you master your world. Through Word Alchemy, you dissolve linguistic parasites and reclaim the creative authority of the Divine Word within.

"In the beginning was the Word… and the Word is creating through us."

Our subconscious is faithfully delivering your heart's desire based on the feeling state your thoughts create. The subconscious takes you literally. Language is the symbol set for your thinking. The subjective mind is a loyal servant; it treats your habitual thoughts, words, and actions as prayers, desires, and commands, then expands them.

Every thought plants the seeds of heaven or hell. Now that you know this, no more mixing signals with scrambled words. We were taught that wise ones think five times before speaking. No more speaking with a **forked tongue.** Say what you mean, act and live on purpose.

The subconscious doesn't "speak English"; it perceives **emotion and pattern**. It translates your feelings into symbols and sorts them by tone and energy level. Intense emotions are prioritized and recalled instantly. Since the subconscious has no sense of time, a triggered memory feels like **the present**. That's why unprocessed experiences from the past replay in the body as if they are in the now.

Cells thrive when the inner environment is stable and conducive to growth. In protection (fight-or-flight), immune function and digestion slow down; blood leaves the forebrain to flow to the muscles. If emotionally charged **words and thoughts** keep firing the survival signal, the body keeps replaying the crisis. Mind is the first medicine and the first illness. Your thinking shapes your state of being.

Many common phrases are abstract to the subconscious and get filed by their most substantial emotional anchor.

Example: "I do **not** want to feel bad anymore." Your conscious mind understands the intent; your subconscious locks onto the feeling of badness and serves it up more. "Do" and "not" cancel each other; "want" vibrates desire without action. Upgrade the syntax and the **signal** upgrades.

Let go of blame. It's about power. You're gaining the tools to starve linguistic parasites and feed the original self.

- Choose wisely.

- Become a spell-breaker.

- Whatever you dwell on is the temple where you pray.

- You are a genie doing daily magic, for better or better.

Create language from your highest desires, not from old limiting thoughts. At first, it takes focus, like learning to walk or drive. Then it becomes automatic.

Practice **being grateful in advance**. Feel as if your desire has already happened. When the feeling is full, the conscious mind, subconscious, and superconscious collaborate. Many of us were trained to pray from a place of lack. Begin to speak using a **creator syntax, use words of a winner,** clear statements that reprogram the mind, interrupt old speech patterns, wire new networks, and move us toward our goals.

The Mimesis Paradox

Mimesis refers to the act of representing and imitating reality through mimicry. It is connected to the term *"meme,"* which refers to an element of culture or a behavior pattern transmitted from one individual to another through imitation.
A meme is an idea, habit, or action that becomes accepted within a group and spreads through repetition. People naturally imitate and conform to the behaviors of their family, tribe, or culture to learn and survive socially.

Myron Golden calls this a **cultural hypnotic societal mechanism,** an invisible frequency influencing our thoughts, beliefs, and actions without our conscious consent.

The effect of these patterns depends on the group's belief systems. Behaviors can uplift or destroy. When leadership is strong and rooted in integrity, and when the group's core values emphasize truth, cooperation, and balance, imitation strengthens the whole.

Nature shows both sides. Mimicry can create order or chaos. Bees and ants, among Earth's most successful species, use imitation to synchronize efforts for the good of the hive.

The Double-Edge of Imitation

Mimicry is not the enemy; it's a tool. It can organize people constructively or corral them destructively.

- **Bright side (order):** Ants, bees, and thriving communities model cooperation, discipline, and shared purpose.

- **Biomimicry (innovation):** Nature-inspired design, Velcro modeled after burrs, and aircraft shaped by birds prove that copying *wisdom* evolves civilization.

- **Neural echo:** The mirror-neuron theory reminds us that *cells that fire together wire together.* We literally rehearse one another.

The 100th Monkey effect

In the 1950s, researchers observed a young Japanese monkey washing sweet potatoes. Something no other monkey had done. Her family copied her; then her peers did. When the behavior reached a specific number, roughly a hundred monkeys on other, unconnected islands began doing the same.

Whether literal or symbolic, it illustrates a **collective consciousness,** a cosmic ethernet linking all life.
When enough humans act consciously, awareness jumps species-wide.

Once enough minds adopt a higher pattern, the field ripples. Critical mass triggers evolution. That is how revolutions of spirit begin.

Mimesis Mutation

Imitation turns toxic when we copy what harms survival: shallow trends, fear mantras, and low-frequency habits. That's the modern dilemma: a culture copying itself into collapse.

Today, many worship **artificial intelligence** as the new oracle. AI can be a brilliant tool, but when we surrender our intuition and creativity to soulless code, it becomes a **parasite**.
People who outsource their thinking lose critical reasoning and imagination. Studies show habitual AI dependence reshapes the brain, reducing focus, memory, and depth while inflating ego and overconfidence.

Wise ones use technology as an **assistant**, not an **authority**.
Keep your organic genius in the driver's seat. If you carry sa-

cred knowledge or rare gifts, share wisely. Seek those with whom you resonate, or be prepared to walk alone.

Remember: **the World is not the Earth.** Each person inhabits a private worldview, an interior planet shaped by their mental wiring. When people say, "The World is ending," they're usually describing *their* world. Words are programs; every repetition is a command. Speak with care, your language can empower or devour you.

Linguistic Parasites

Let's review what a parasite is again. A parasite needs a host; it alters the host's behavior, consumes the host's energy and resources, and uses the host to move from one host to another for reproduction or duplication. This precisely describes what language parasites do: they hijack experiences, manipulate emotions, perspectives, dictate actions, and create contracts. They assign labels and shape our reality based on belief systems that may lack a factual basis.

Words play a significant role in how we experience life. We tend to mimic common sayings or memes that are repeated across cultures. Whether or not they are socially accepted, the subconscious treats such statements as literal orders. We must be cautious not to echo phrases that keep us in mental bondage, feeding the parasite. These sayings are so normalized that people don't even realize they are hypnotic triggers, small commands shaping behavior, identity, and expectation. Words do matter.

Here are some Popular meme statements and Slogans used in North America to be aware of:

You must take the shot to protect others.

Time is money.

80% of businesses fail in the first …

Hard work always pays off.

It's just the way it is.

It's always been this way.

You know what they say….

That's just the way we are.

That's just the way it is.

You know how they are.

Good things never last.

We are living in the end times.

You can't change people.

You can't change things.

It's in your DNA.

It runs in your family.

Fruit doesn't fall far from the tree.

You're before your time.

You can't change the system.

One monkey don't stop no show. (The 100th monkey experiment disproves this)

You can't believe anything.

I'll believe it when I see it.

You can't trust anybody.

You must use AI or you'll be left behind.

Each of these statements carries a **vibration** that either affirms limitation or unconsciously seeds fear, exhaustion, or scarcity. When repeated, they become programming codes that govern our emotional tone and biological state.

We must think critically and consciously when we speak, while being aware of the toxic communication patterns of domesticated **sheeple**. Sheeple exist in survival mode, blindly repeating phrases that hypnotize the herd. This linguistic mimicry is one of the most powerful tools of mass persuasion, a marketing tactic that disempowers individuals, reshapes neural pathways, creates chronic consumers, and disconnects people from their original purpose for being.

Every word carries a frequency, a tone, and a molecular signature that travels through your body and the bodies of those who hear it.

When you speak, you vibrate water. Your body is over 70% water, a conductor of emotion and energy. Every spoken word, thought, or affirmation literally reorganizes the geometry of the water within you. Masaru Emoto's research proved that water forms harmonious crystals when exposed to words like *love* and *gratitude,* and chaotic patterns when exposed to words like *hate* or *fear.*

Your cells listen. Each thought you repeat becomes an *acoustic order* to our bodies. Our DNA is not static; it's a **liquid-crystal bio-antenna**, responding to sound, emotion, and intention. This is not a metaphor; it's physics.

When you say, "I'm sick," your cells prepare for illness.
When you say, "I'm done," your nervous system begins to shut down.

When you say, "I can't," your subconscious locks the door to possibility.
But when you declare, "I am healing," or "I am rising," your cells light up with coherence. They start producing peptides, hormones, and electrical signals aligned with that command.

Every word is a spell, that's why it's called *spelling.*
You're either casting curses or creating cures with every breath.

Words are the bridge between thought and manifestation, the invisible software that programs reality. When you change your language, you alter your frequency. When your frequency shifts, your results shift.

Language can be used to create opposition or liberation, life or death. Words definitely do matter.

From Emoji to AI

Writing this book challenged me. I was labeled dyslexic as a child. Letters used to dance off pages. I'm an auditory, kinesthetic learner; words live in rhythm and movement for me. At one point, I asked AI to polish a section. Seconds later, it looked sharp, but when I ran a plagiarism scan, the tool admitted it had **copied** from multiple sources. And here's the twist it told me *I couldn't legally own its words.*
That was the moment I saw how easily the machine mimics without soul.

The new mantra: *Use AI or be left behind.* But as Nipsey Hussle said, blind competition kills creativity. When we chase the algorithm, we feed the parasite.

We spend hours staring into glowing rectangles, letting devices sculpt our minds. Texting reduces communication to fragments. Then came emojis of round yellow faces that flatten real emotion. Symbols become shortcuts for feelings we no longer articulate. Yellow itself carries vibration: joy, intellect, but also caution and cowardice, the color of the solar-plexus chakra, our willpower center. When that energy wheel spins weakly, we feel stuck, fearful, and ashamed. Even our icons reveal imbalance.

Each symbol carries an **agreement,** a micro-spell training the collective mind. Slowly, we're losing nuance and authenticity, like cattle fattened for slaughter. As the alien Faraday says in *The Man Who Fell to Earth:*

> "The misperception regarding communication on
> this planet is the illusion that it has taken place."

Screens emit frequencies that alter brain chemistry. Blue light damages retinas, suppresses melatonin, and keeps the mind restless for hours. It accelerates aging, anxiety, insomnia, and even cancer. I use blue-light filters, grayscale mode, and EMF-harmonizing devices from **improveyourlife.us** (tell them Doctah B sent you). These don't block radiation, they **shift frequency**.

We are drowning in data yet starving for wisdom. This is **Artificial Ignorance,** knowing everything and understanding nothing. The flood of input disconnects us from the authentic self. We built electronic superbugs that drain life force, turning users into self-replicating zombies. Hence the name: **Cell Phone** because it's both *cellular* and *a cell.*

Simple remedies:

- Take digital fasts. Use "Screen Time" to monitor usage.

- Power down 30 minutes before bed; keep devices at least 6 feet away.

- Ground yourself outdoors. Walk, breathe, talk face-to-face.

- Read real books. Write by hand. Reconnect to analog soul.

Our great-grandparents ate foods alive with prana grown in wild soil, not factory labs. Today, we consume lifeless anti-foods that dull intuition and shorten lifespan. Modern humans have agreed to live like livestock, domesticated, herded, and fattened for market.

Those who live differently are mocked or attacked for independent thought. Memes persuade the masses to abandon individuality. That's how the **Watiko Virus** feeds, devouring creativity, spirit, and vitality. It is the psychic pathogen behind **colony-collapse disorder** in the human hive: a virus of mass distraction. Seeking external validation drains genius; it breeds passivity the herd waiting to be consumed.

The paradox is apparent: **mimicry can heal or harm.** Choose to imitate wisdom, not weakness. Use the mimesis effect consciously to upgrade, not downgrade, humanity.

Consistent Inconsistencies Cause Chronic Confusion

Consistent inconsistencies create collective madness. We are bombarded with mixed signals that contradict nature and common sense. When the mind is forced to hold opposites too long, it fractures.

Our essence strives to heal, grow, and ascend, yet we're spoon-fed lies. Overexposure dulls discernment. Soon, confusion feels like home.

Parents plant the first seeds, Santa Claus, and the Tooth Fairy fantasies that condition children to accept untruth wrapped in love. Later, adults wonder why honesty feels foreign.

Holidays we celebrate mindlessly were once **esoteric codes** for cosmic cycles. Now they're corporate rituals of debt and dopamine. Autosuggestion repetition turns myth into habit. Western culture is addicted to ritualized contradiction.

Each year, we relive the same chaos, spending, stressing, and striving, calling it tradition. These patterns, passed down through church, school, and media, promotes **chronic servitude syndrome.** Life force becomes food for unseen parasites.

Inconsistency breeds disease. When reality clashes with truth, immunity falters. In chaos, the body stays in fight-or-flight; in love, it finally repairs.

Fear-based media monetizes alarm. Every headline screams danger. This constant stress loop fuels illness. Positive events are buried by design. Inconsistency becomes infection, replicated through every broadcast.

When perception aligns with love, peace, health, and worthiness, the body's biochemistry mirrors that harmony. The subconscious builds cells from your dominant thoughts. You're always coding yourself by default or by design.

Epigenetics and Pattern Interrupts

Epigenetics proves that the environment shapes gene expression. Food, behavior, toxins, and emotions all sculpt DNA activity. Change the field, and you change the fate.

Use **pattern interrupts,** subconscious spell-breakers to shatter loops. The body self-repairs when nourished with clean food and pure thought. Unplug. Silence the noise, fast from fear.

A.N.T.s Automatic Negative Thoughts

Not sure where I heard this statement, but it is valid. Are you being eaten by **A.N.T.s**?
Automatic Negative Thoughts swarm the mind like parasites,

idea pirates hijacking your mission. They tighten their grip each time you resist.

Persistent self-conflict shortens **telomeres**, the protective ends of your genes. They're like antennas, transmitting your emotional environment to your DNA. If your daily signal is hopelessness, cells receive the order to self-destruct. This is how chronic stress accelerates aging.

Your thoughts are leadership commands to the cellular crew. Speak chaos, and the crew abandons ship. Speak life, and they rebuild the vessel.

They Live. A Warning About Perception Management

Mainstream media casts a broad net, colors, jingles, and edits to hook attention and **reprogram neural networks**. Viewers mistake storytelling for truth. This is **perception engineering**, selling belief as fact.

Commercial-free alternatives exist, but herd mentality defends the mainstream. Anything outside the algorithm is labeled a *conspiracy.* That's control by ridicule.

In 1988, John Carpenter's *They Live* exposed the machinery of control, mindless consumerism, subliminal marketing, and psychic hypnosis. Billboards that appeared to sell convenience actually hid commands: **OBEY. CONSUME. CONFORM. SUBMIT. BUY. REPRODUCE.** Citizens walked through life under an enchantment, unknowingly feeding alien overlords of thought and appetite.

A blind minister creates glasses that reveal truth: the aliens' grotesque faces and the secret words behind every ad. The film's most extended scene, a brutal fight between friends, shows how fiercely humans resist seeing reality.

People cling to conditioning even when evidence screams otherwise. Entire wars are fought over defended illusions.
So be wise in how you share revelation: some aren't ready to see. **Conserve your life force. Find or build your tribe.**

In *CA$HVERTISISING*, Drew Whitman outlines eight primal human drives. Two are *Freedom from Fear, Pain, and Danger,* and *Care and Protection of Loved Ones.* These biological impulses often turn people into zealots, trying to force others to change. Be cautious such crusades can get you ostracized or worse.

To free the mind, create **new habits.** Practice new behaviors until neurons re-link. Pair **Subconscious Alignment** with the **28-Day Elevated TotalBody Paracleanse** and watch the quantum shift. You'll feel lighter because what's been eating you mentally and physically is gone.

The Issue Is in the Tissue

The subconscious lives in every cell the **body-mind**, not just the brain. Many illnesses are crystallized emotions. Eastern systems knew this long ago.

- **Liver** → **Anger**

- **Spleen/Stomach** → **Worry**

- **Lungs** → **Grief**

Release the emotion, and tissue renewal massage, breath-work, and movement, free trapped energy. Tears are detox.

Holding a feeling of unworthiness for years, it becomes flesh. Practice worthiness for 28 days, and the parasite dissolves, vaporizes, neutralizes, and transmutes. The body reorganizes around the new vibration. Healing that looked miraculous is simply alignment restored.

Healthy thought amplifies healthy cells. Toxic food breeds toxic energy. We are what we absorb nutritionally, emotion-ally, and energetically. Claim mastery of your terrain. Eradi-cate what's eating you from the inside out. The body is de-signed to store **love, peace, and harmony** if that's the domi-nant signal.

Reclaiming the Wise Ways of Old Days

Nature is a sacred orchestra of polyrhythms and harmonies. Ancient people moved with that music, seasonal, cyclical, balanced. They ate what grew nearby, when it grew. That synchrony bred longevity.
Change the food, and you change the people.

The Industrial Revolution introduced synthetic food and syn-thetic desire. The buy-now-pay-later era birthed diseases of excess. Before that, food was local, alive, and mineral-rich.

Today, a few **blue zones** still remember. They live, commu-nally, intentionally, as centenarians who thrive, not just sur-vive.

Indigenous peoples understood that all things possess living essence, **Animism**, the root of original spirituality. They healed through ceremony, fasting, song, and alignment with the land.

When conquerors came, many were infected not just by disease but by ideology. Yet remnants of those ways still survive, whispering through us now.

Modern society, addicted to stress and spectacle, has normalized sickness. Only an unwell civilization would celebrate destruction, distraction, and death.

We are witnessing the **Watiko Virus** the psychic pathogen of our age creating colony collapse in the human species. But this invasion only succeeds if we allow it.

You are holding the antidote in your hands.

To reclaim your mind, body, and spirit, merge the **ancient ways** with **modern wisdom** and forge a new way of living or cease to exist. We stand at the threshold of the Sixth Mass Extinction. The only way forward is inward.

Natural Solutions to Living Your Best Life Now

> *"Victorious warriors win first and then go to war, while defeated warriors go to war first and then seek to win."*
> **Sun Tzu, *The Art of War***

Make up your mind that victory is certain.

Another timeless truth from Sun Tzu reminds us:

> *"If you know the enemy and know yourself, you need not fear the result of a hundred battles."*

These teachings apply perfectly to the inner war we all face. The *enemy* today is not always external it often lives in the form of **self-limiting habits**, **mental parasites**, and **energetic addictions** that eat away at our peace, purpose, and power from the inside out. The most powerful enemy to be conquered is within ourselves.

To win this war, we must first **acknowledge our own patterns**, understand our inner programming, and consciously choose transformation. Self-awareness is the first sword of victory.

Chapter 18:
Subconscious Alignment Protocol

"When the conscious and subconscious agree, reality, we become the captain of the journey of our lives."

The Inner Reset System

Doctah B's Subconscious Alignment Protocol is a revolutionary system designed to help rewrite undesirable mental programming, control unhealthy habits, undo unwanted social conditioning, and assist in preventing and healing dis-ease on every level.

This sacred process clears psycho-energetic blockages, dissolves emotional barriers, and awakens the natural experience of vibrant health, abundance, and success. It's a direct upgrade for your inner operating system, aligning you with the blueprint of your original, unlimited self.

This method draws from a unified field of disciplines: Epigenetics, Quantum Physics, Meditation, Ayurveda, Psychological Kinesiology, Neuro-Linguistic Programming (NLP), Cosmic Resonance Theory, Subconscious Mind Theory, and the Laws of Attraction. Together, these sciences reveal that our thoughts, emotions, and energetic frequencies can re-encode the body and reshape reality.

- Epigenetics: The science showing how environment and perception can turn genes on or off, proving that outcomes are programmable.

- Muscle Testing (Kinesiology) – A biofeedback method that communicates directly with the subconscious through the body's natural muscle/ motor responses.

- Cosmic Resonance Theory: The concept that vibration and frequency organize matter and experience, meaning your resonance determines your reality.

Body Language

Over ninety percent of all communication happens beyond words. The subconscious mind governs all motor functioning and the nonverbal field we call body language.

Because of this, a specialized form of Muscle Testing is used to communicate directly with the subconscious. When done correctly, it becomes an accurate mirror of hidden programs and beliefs shaping your experience. Since the subconscious controls bodily responses, we can literally ask the body questions, and it answers. This allows us to identify and transform limiting beliefs, emotional blocks, and outdated conditioning that no longer serve our evolution.

The process is natural, noninvasive, and profoundly effective. Through guided subconscious dialogue, vibrational recalibration, and focused intention, the client is gently led through a sacred inner gateway to a serene, radiant space deep within the mind, where the infinite intelligence of the true self awaits. From this state of heightened coherence, a series of alignment exercises is performed to reprogram the mind toward a desired outcome, awakening the life you were born to live.

Transformational Outcomes

Participants frequently describe becoming more proactive, creative, and success-oriented, which has led to greater clarity, confidence, and personal power. Many report:

- Attracting new opportunities and prosperity

- Releasing self-sabotaging beliefs and behaviors

- Restoring health, vitality, and emotional balance

- Achieving conscious-subconscious harmony (mental synchronization)

The subconscious is not your enemy; it's your greatest ally.

The conscious mind is the seed; the subconscious mind is the soil that nurtures it. If the seed is weak or the soil is toxic, the plant struggles to thrive. When the seed is vibrant and the soil fertile, the plant grows strong and bears abundant fruit. Subconscious Alignment is the art of tending your inner garden, pulling weeds, enriching the soil, and consciously cultivating the harvest you desire.

You Are the Authority

Throughout every session, you remain entirely in control. Your Higher Self, your inner wisdom, is the ultimate guide, determining which beliefs are ready for transformation. You receive the golden keys to operate the most advanced supercomputer ever designed: the Subconscious Mind.

This alignment gently returns you to your natural path, allowing deep healing and a personal paradigm shift to unfold naturally, organically, and spontaneously.

Mindfulness, breath, and theta practices help dissolve old ego states and write new programs. The target is balanced mind-body, present-time awareness, less emotional reactivity, and more flow. In that state, new instructions sink in more deeply with less resistance. That's how mental parasites lose their grip.

Smart Devices: Use the Tool, Don't Become a Tool

There's a modern pattern of **data-driven dissociation,** long hours of scrolling, and prompts that keep the nervous system buzzing and the spirit distracted. Handhelds can become **electronic drugs** if you let them. Notifications and engineered prompts can train feelings, thoughts, and reactions minute by minute. That's attention poverty and energy leakage.

Here's the empowerment move:

- **Use the device as a tool, not a teacher.**

- **Phone fast** like you would food fast.

- **Decide** when to engage don't let algorithms decide for you.

If you've ever lost your phone and felt a jolt, that's honest biofeedback. I've felt it too. The fix is simple and strong: scheduled **device fasts**, silence windows, and **notification auditing**. The nervous system resets: the original rhythm returns.

Be careful of sensational facts traps, history-lite narratives, made-for-clicks news, and influencer scripts. Without active discernment, external stories can erode intuition. The antidote is **independent thinking** and **deliberate inputs**. Carefully curate your feed and your food.

Be aware that smart tech is everywhere TVs, cars, meters, speakers, locks.

- Check privacy, security, and **location** settings turn off what you don't need.

- Ask utilities about **opt-outs** or health accommodations for smart meters; some regions honor them with proper documentation.

- Use a **VPN** and privacy-first browsers.

- Schedule **power-down nights** (breaker off) or at least router off; notice how it affects your sleep.

- Track your actual **screen time**, then shave it down by design.

The point isn't fear. It's a **choice**. Use the tools; don't let them use you.

Sex Alchemy

Sexual energy is a totally creative vibration. **Conscious coupling** can be used to give birth to ideas when two people share a precise, loving chief aim. The goal could be to create more love, peace, finances, and better health. Whatever you are thinking and feeling during this sacred connection, you are attracting and giving birth to. In this love zone, the body releases **sacred secretions,** a special mix of hormones and energies programmed by your combined mindsets. This is called sexual transmutation.

Imagine speaking to your genie while rubbing the lamp: "What is your desire?" Feel as if your desires have already happened.

Unconscious coupling lust, violence, trauma-bonding, or partnering with someone hosting heavy mental/physical parasites cross-contaminates the energy field. The result is more opposition in life, not fulfillment: This keeps both parties living in the "realm of the hungry ghost." Be haunted by discord and drama. Repeating that loop drops us into the lower animal mind and feeds old habits that eat us alive.

Flip the script:

- Try this mantra: We are magnetic, attracting good health, moor wealth, and pure love.

- Make love **out of** what isn't love. That's sex alchemy.

Messages In The Music

Music is a universal language that can be used as medicine or poison, as a liberator or a parasite. Imagine life without it. It's hard to do. The universe is a **symphony of vibrations,** endless rhythms, and melodies. What sets one thing apart from another is the **song** it sings.

Music transcends language barriers and reaches both the conscious and subconscious mind. The person who creates the soundtrack can influence moods and feelings. It can bring harmony or be used as a weapon.

As musicians, we communicate without words. We can meet on stage and improvise for hours. My heart beats music that turns chaos into order, and creates love out of what isn't love. Many genres can uplift us; improvisational jazz and indigenous world music are meant to do just that. Music induces a trance and meditation, engaging both the conscious and subconscious minds simultaneously. It influences gene expression through the chemistry of emotion. Music can organize or scatter the mind; it is a powerful psychoactive substance.

Most mainstream commercial soundtracks today are harmful parasitic machines designed to induce disease, hypnosis, nervous system imbalances, destruction, depression, disharmony, materialism, fear, pain, greed, obedience, separation, consumption, and outright evil. Much of the chart-topping music today is based on trance-like loops that encourage hypersexuality, illicit money, secret rituals, and drama, which creates more wealth for the pharmaceutical, prison, and alcohol industries. To add insult to injury, these projects are tuned to the concert pitch of A 440Hz, which, in my experience, causes physical and psychological stress.

Like food, the quality of music directly influences our health and can shape our thinking on the deepest levels. For most cultures, music molds reality and is fundamental. Remember, it is the repetition of pleasure or pain that programs the subconscious mind, and music can induce both states.

For over 50 years, social engineers have completely controlled the entertainment industry, also known as the enter-train-ment industry. These oppressive power structure elites use entertainment as a tool for crowd control, shaping a culture of mindless, domesticated followers who will follow leaders off a cliff without hesitation.

To succeed in this industry, artists, writers, producers, actors, and musicians are forced to submit to the will of demons or face starvation, or even worse. Oh, by the way, we artists, writers, and producers have never been fairly compensated for our creations; we have been slaves working hard to feed the parasitic slave masters. The good news is the old music industry model is all but dead, and I got out just in time.

True creative artists are revolutionaries, cultural creatives, inducing change, liberating minds, and creating melodies and rhythms in the most beautiful of ways. Direct distribution makes it possible to share truth without feeding the old parasitic music business monster that was eating artists alive, while promoting opposition and feeding the Watiko mindset.

I have created a project called **Doctah B, Medicine Music Collective Special Edition**, as a soundtrack for this book. Medicine Music is curated and designed to balance the neurological system. It features rich soundscapes tuned to natural resonances like **432 Hz** and **528 Hz**. These are blended with sounds of nature, healing rhythms, indigenous and traditional instruments, analog warmth, and digital clarity. We incorporate **binaural** and **trinaural** patterns with **theta** waves to promote whole-brain states, creativity, meditation, and pleasure. I have been using **the musical** frequency-based subconscious alignment protocol since the year 2000. The First project was called Herban Shamen, created by Akadamah Jackson and me as a healing modality.

My goal is to create sound environments where mind, body, and spirit synchronize and reduce stress. This raises life-force energy, increasing vitality so we can rise above what's been eating us, creating Total transformation.

Your New Daily Rhythm

- **Speak only of what you desire. Keep It Simple.**

- **Feel first.** Gratitude in advance flips the signal from lack to creation.

- **Practice** breathing deeply while imagining your desires as if they have already happened; This reprograms the mind.

- **Fast from noise.** Phone fasts, news fasts, power-down nights, and disagreeable people fast.

- **Curate your life.** Music, media, mentors, choose what fuels your purpose.

- **Transmute.** Aim sexual energy at healing and high outcomes.

- **Protect privacy.** Tighten settings on phones, tablets, watches, and TVs, use a VPN, and schedule no-screen hours. Be aware of what you're saying around smart devices; they are listening.

- **Repetition is key;** it creates the spell rhythm. Make sure your daily thoughts, words, and actions are moving you from opposition to liberation.

You are not broken; You are a genius, brilliant software running outdated code. Update the program. Align your words, thoughts, feelings, and actions. You're the Genie, your genes obey your commands; your world responds to your vibes faithfully.

Chapter 19:
Wake Up And Break Free

If you're reading this, the healing has already begun.
Awareness itself begins to starve the mental, physical, and energetic parasites that feed on confusion, discord, stagnation, and complacency.

In 2003, Vashti Bonner had a vision. At the time, I was feeling drained and lost, only giving small local lectures for offerings, basically selling herbs out of the back of my car, hoping the World would hear my message. She proposed an extraordinary idea of creating an educational oasis. This new energy woke me up and helped me become aware of what was eating me, a parasite called stagnation.

We teamed up, and she curated a beautiful natural health boutique called Elevation Foundation in Los Angeles while offering me much-needed inspiration and guidance. Each week, people came from far and near to be part of the Elevation Experience. We offered an exclusive line of natural products, herbal supplements, crystal jewelry, natural fiber clothing, rare resins, fountains, tonics, and elixirs. We provided classes, workshops, and lectures to educate and elevate minds. In the back of Elevation, there was a herb garden area where we would host events. When the weather was favorable, I would conduct lectures, subconscious alignment sessions, classes, and health consultations out there. Often, clients would remove their shoes and ground themselves while receiving sub-conscious alignments. The combination of touching the earth and being outside under the sky creates a natural energy flow that can help open blockages in our auric field and chakra system. This made the experiences much more profound.

There was one client who had a history of generational and personal trauma. She came to see us many times, and no matter which alignment protocols or balances were used, they had no effect. On one particular session, totally frustrated, she cried and fell asleep. In the garden, there were always bees and butterflies around. I watched them as if looking for an-swers, and I drifted into a kind of dream state. Suddenly, my mind was full of new information. I awakened the client and asked her to try something.

I created a musical instrument called the **Medicine Drum, tuned to a 432Hz healing frequency**, that I would play during presentations to help awaken the theta brain state, where im-agination is quick to arise. I instinctively began playing a 6/8 rhythm on the drum while taking her on a guided meditation journey to a place inside her where boxes or containers of old experiences, emotions, and memories lay deep in her foun-dation. I guided her to remove these boxes one by one, with-out opening them, and place them on an altar, all in her mind. The special medicine, music, and guided meditation created an alternate dimension where she was able to non-emotion-ally incinerate epigenetic traumas without having to locate or relive the issues.

The old ways of therapy have been known to retraumatize some people. Next, I guided her through a set of 7 movements that helped reset her nervous system and open a new window of infinite possibilities for her. When we were done, she started screaming like crazy and jumping all over the place. At first, I thought, "Oh shit," she might have snapped and lost her mind. Then suddenly she just ran off, crying, jumped in her car, and sped off. I sat there confused about what had happened to her and where this whole process had come from. It was as if I were guided from another realm, which occurs at times, especially when I'm doing lectures or creating music.

I sat there in the garden for some time, feeling emotional and going over in my mind this process and the events that had transpired. Then something said to look at my phone, which I had turned the ringer off during the session. The client had called several times and sent pictures of herself at the beach, on a rock, with her hands high in the air, along with several texts saying she felt healed, renewed, and awakened from a deep sleep. I called her, and she said that the session was the most magical experience she had ever had.

She said it was like years of therapy in one session. She asked what it was, and I said I'm not sure; something came over me. She said, "Well, I'm glad it did, because she was having a huge breakthrough and had finally ended." I got very excited and ran into the elevation office, asked the people there to let me try something on them, and they saw my excitement and said, "ok." I took them through the process, and afterward, they were emotional and repeated themselves. What was that? Out of my mouth just blurted out The Foundation Balance.

Since then, we have done this with individuals and in groups with astounding results each time. Over the years, we have refined the process and rebranded it as **The Core Access Experience** because it achieves access to the subconscious core, where many of our challenges are stored in containers ready to be incinerated. It is a challenge to describe what happens during the process, but this touches the surface.

The Core Access Experience Explained

When you or your group is ready to go beyond understanding and step into real transformation, the **Core Access Experience** is the next evolution of your healing journey.

This live process is a fusion of subconscious alignment, breathwork, movement, sound healing, and guided meditation designed to **reset the inner architecture of your mind and body** and reach the subconscious core.

During the experience, participants are guided to create a powerful **Core Goal Statement,** a new intention programmed directly into the subconscious, then gently led through a live **432 Hz Medicine Music Soundtrack**, rhythmic breath, and mindful movement that carries the brain into the *theta state*, where deep creativity, intuition, healing, and rewiring occur.

The process integrates **pranic healing (Vital Energy), quantum visualization, and a seven-directions ceremony** to harmonize the energy field and seal the transformation with a recallable "core-anchoring" technique. The result is a new dominant goal, a clear mind, lighter emotions, renewed focus, and the ability to activate the core access experience state on command.

Many describe Core Access as *ten years of therapy, coaching, and spiritual practice condensed into one sacred evening* a direct experience of freedom from the mental, emotional, and energetic parasites explored throughout this book. If you or your group is interested, then inquire at **www.elevation-time.com**

Are you ready to move from opposition to liberation and break free from parasites?

No one is coming to save you! You must commit wholeheartedly to investing your own energy in transforming your life, time for you to save you. This decision must come from *you*, not from fear or outside influences.
Liberation begins the moment you say **yes** to your own evolution, to your elevation, to your liberation. Let's go!

Elevated Action Steps

1. Reset the Body and Mind

Fast intentionally.
Give your body and mind time to rest and repair. Fasting cleanses the system and resets your frequency for healing and mental spiritual clarity.

Nourish the temple.
Take natural supplements that support vitality and regeneration.
(Explore premium formulas at ElevationTime.com.)

Eat to live. Decide to stop committing Nutritional Suicide.
Choose living foods that energize, vitalize, and rebuild you. Avoid processed, lifeless foods that feed both physical, mental, and emotional parasites.
Once you cleanse, healthy cravings naturally take over.

Cleanse seasonally.
Commit to a **28-day systemic parasite cleanse and rejuvenation program four times a year,** one per season, to stay in rhythm with nature's cycles.

2. Reclaim Your Inner Space

Use your imagination.
Let yourself envision what you truly desire without the opinions, ideas, beliefs, or judgment of others.

Create sacred time.
Even **7 minutes a day for 28 days** of deep breathing, visu-

alization, or meditation can rewire your nervous system for peace and harmony.

Go on a media fast.
Avoid the news for seven days.
The news is like a noose; it chokes the mind. If something truly matters, you'll hear about it naturally.

Break the spell of possession.
Let go of the people, habits, words, thoughts, and beliefs that drain your light. People who are possessed will try the hardest to possess you.

3. Recognize the Parasite Voice

The parasite voice speaks through fear, guilt, gossip, lack, dissatisfaction, blame, self-doubt, excuses, judgment, constant comparing, belittlement, deflection, and the disagreeable past.
It disguises itself as logic, realism, or "just how it is."
Learn to notice them, become the hunter, and track down and eradicate the parasites that have been eating you alive.

4. Self-Inventory and Soul Alignment

List your assets.
Write down the strengths that elevate your life.

List your liabilities.
Acknowledge the habits, attitudes, or patterns that weaken you.

Audit your circle.
Who speaks 80% solutions, and who speaks 80% problems? Energy is contagious.
Align with people whose vibration matches or exceeds yours. Never downgrade your energy to make others feel comfortable. Never.

5. Use your powerful imagination. Embody the New You.

Feel the new you in your body now.
Hear your new voice, clear, calm, confident.
See your new life through new eyes.
Fall deeply in love with the version of you that's emerging.
You've got the whole world in your hands, just imagine.

Mantra:
I forgive myself, my family, and others for imperfections,

Create a Dreamtime Meditation.
At bedtime, as you're lying in bed, repeat your new goal statement mantra over and over silently every night as you drift off to dreamtime.

Repeat it each morning upon waking. Before you step out of bed, draw an imaginary circle on the ground. Mentally place your new definite chief goal statement into the circle. Then step into the circle. This plants the new idea seeds into the fertile soil of your subconscious mind.

All while you're in dreamtime, your mind waters, nurtures, and plants the seeds of your thoughts. They are guaranteed to grow fruit.

Repetition is sacred technology; it reprograms the sub-conscious for better or for worse. It's all up to you.

6. Gratitude: Create Heaven on Earth

No more begging prayers, beggars can't be choosy. Be grateful **in advance. Live as if the dream has been ful-filled, as Nevile Goddard teaches.**
Gratitude collapses time; It moves you into the vibration of completion where *Time equals Art, not Money.*

This elevated state boosts immunity, speeds healing, and magnetizes success. Gratitude collapses time.

7. Watch Your Words and Thoughts

Catch yourself when you replay old pain or retell the story of struggle, failure, blame, and defeat.
Each time the past tries to speak, recognize it as an old-time echo, then replace it with your new command. **You are the author now.**

8. Reprogram the Mind

Seek practices that align your **conscious and subcon-scious** breathwork, meditation, subconscious alignment ses-sions, sound healing, or movement arts.
There's no single "perfect" method. The right one is the one that truly works for you.

(Explore our time-tested protocols and formulas designed for this process at www.elevationtime.com.)

9. Daily Visualization Practice

Take time each day to imagine **your desired reality:**

- If healing is your goal, **feel** wellness now.

- If abundance is your goal, **feel** prosperity now.

- If peace is your goal, **breathe** it into every cell.

The subconscious obeys the **feeling**.
Become full of the feeling, not just the words.

10. Patience, Persistence, Power

Transformation follows rhythm:
28 days to plant new habits.
90 days to anchor a new reality.

Be patient. Be consistent. Be well

Mantra: I allow myself to be healthy, wealthy, and wise while inspiring others to do the same.

11. Elevate Your Journey Join Us

Join us for **wellness retreats, workshops, and classes** designed to help you awaken your highest potential and create Heaven on Earth from within.

It's our mission to guide you through this process of liberation. This book, *What's Eating You?,* is more than knowledge. It's a roadmap to freedom from mental, physical, and energetic parasites.

We've shared the tools. The keys are in your hands.

Are you ready to break free from what's eating you and create Heaven on Earth? Imagine it happening inside you, right now?

Chapter 20:
The Daze Of The Weak!

We are all born with unique abilities, but only genuine seekers recognize, honor, and cultivate them. In today's world, individuality is often treated like a disease. Those who dare to be different are labeled weird, crazy, autistic, too smart, or "too deep."

Remember the question: **"Who do you think you are?"**
If you don't know, the world will decide for you.
Modern culture glorifies "normal," rewards conformity, and fears originality. Many have agreed to be spellbound, sheeple just counting the **Daze of the Weak,** sleepwalking through a **daymare** while wearing masks just to fit in.

No more eating food that eats you. No more following the script written for you. No more living in a prison of social norms that kept you in bondage. No more self-betrayal that leads to a world of self-repression, self-manipulation, allowing your mind to devour itself. The slow death of authenticity is one of the root causes of modern illness. It feeds psychic infections, mental parasites, Wetiko, generational curses, and viruses of the mind.

The mind is the first medicine, but it's also the first disease. Never forget this fact.

The Alchemy of Opposition

For the awakened wise ones, opposition becomes fuel for greatness. Parasites, pain, and resistance become teachers' tools for transformation. Used consciously, they strengthen us; left unchecked, they consume us.
This is the seasoning process the alchemical fire that forges the soul.

Every parasite, mental, physical, or energetic, has an Achilles' heel, a blind spot. It can only survive by feeding on its host's weaknesses and ignorance. Once you wake up and choose to be the captain of your vessel, all parasites must abandon ship or die.

Healing is a choice. Liberty is a daily practice. You must rinse and repeat. It takes courage to be transparent, express your true self, embody your genius, and detoxify physically, mentally, emotionally, and spiritually. You must be willing to transmute, eradicate, compost, seek out, and destroy the thoughts, words, and actions that have been eating you alive. It's not always easy but it's simple when you take one step at a time.

Choosing the path called FAITH (verses hope)

I love the way my friend, the amazing Sage **Iyanla Vanzant,** breaks down **FAITH, Feel As If The Thing Has Happened.** Now that's really deep but simple. Those who rise above "normal" become the pioneers of possibility. Remove the safety net, trust your path, and create miracles from Iyanla's version of the metaphysical FAITH.

Once you recognize what's been eating you, you reclaim the power to change everything.

The Polarity Principle, The law of correspondence

Life on Earth comes with challenges: disease, trauma, greed, fear, war, and parasites both seen and unseen.
But all things exist in polarity. All things are connected to their polar opposite. There could be no up without down, no black without white, no light without shadow, no illness without a cure, no good without evil.

The key is perspective. Change your mind, change your life.

Even nature teaches this:

- Diamonds form under pressure.

- The Blue Lotus rises pure from the mud.

- The Moroccan olive tree thrives in hot barren soil.

- The ancient chaparral bush endures the desert sun as one of the oldest living medicines on Earth.

Challenges can either make us stronger or take us out. The choice is ours.

99.9% of the universe is space, a field of infinite possibilities. Our thoughts project into this quantum field as waves of energy. Repeated often enough, those waves collapse into matter. We are co-creating reality with every thought, word, action, and feeling.

Ancient people understood this through *animism,* the knowing that everything is alive, conscious, and connected. They lived in harmony with nature, not in opposition to it. To survive as a species, we must return to that wisdom, or self-destruct by eating ourselves alive.

Chapter 21:
The Inner Revolution

For too long, we've celebrated memorization over imagination and worshiped systems that dull intuition and glorify struggle. The most dangerous enemy we face is not outside; it's inside. In the past, we may have been programmed to eat ourselves to death by parasitic beings, but the time has come for each of us to rise up and put ourselves in the winner's circle at once. No more cannibalism of the soul and begin the alchemy of renewal.

In *The Four Agreements*, Don Miguel Ruiz describes the "parasite" as the collection of false beliefs, fears, and judgments that feed on our life force.
He reminds us that the real battle is internal between the voice of truth and the voice of the parasite. We have the power to choose!

We must teach our children and ourselves to trust creative independence, emotional intelligence, and spiritual awareness. That's how we rewire humanity.

The Hitchhiker Effect

During the journey, opposition to liberation, it's imperative to watch out for a special type of superbug that is a master of disguise. Once you have it, it hides deep within the brain within the limbic system and the RAS (Reticular Activating

System), both of which are essential for survival. This is our control center that processes emotions, memory, and behavior.

Imagine this analogy: an innocent-looking hitchhiker on the side of the road, standing in the rain, who happens to be going the same direction as you. It may be holding up a sign that says it just needs a ride to a place called Liberation.

You may be able to relate to this situation because you are familiar with the feeling of being all alone in the rain, needing to be picked up by someone you can relate with. For a split second, your gut brain says, "Don't trust strangers," and the cranial brain says, "This could be you one day. Be a good neighbor; pull over and let them in."

Once you pick it up, it begins to mirror you and convinces you of how much you have in common with it. During the ride, the two of you start bonding, you realize you know some of the same people and discover you may even be related.

The parasite convinces you that it is very knowledgeable about life and can be totally trusted with your ideas, dreams, and aspirations. By now, you unknowingly allowed it to plant a seed in your mind that causes you to trust others' opinions about your life more than you trust your own. After a while, you may feel a bit tired, so when it asks, "Would you like me to drive?" you say yes. What are friends for?

After some time, you miss the off-ramp to **Liberation City** and decide to ride with your newfound companion on a long journey to Nowheresville.

This mental parasite's food is the vapor you emit when you feel lost, incompetent, never trusting your ideas, drifting through life, feeling the need for constant external validation.

This parasite causes you to distrust yourself and to constantly seek other people's opinions and judgments to guide you on your personal journey to a place they have never been. This superbug is a plague that has handicapped, broken, and consumed the spirits of many natural-born geniuses. It causes people to devalue their own ideas and to trust gut feelings while chronically seeking validation from others. It may be a real challenge to eradicate because the hitchhiker parasite is a mindset that runs in families, locks down communities, and eats the creative spirit. This thing creates a culture of mind-dead people mentality in the grave, just waiting for dirt.

The first step in the eradication process is recognizing when you're talking too much and oversharing. Keep your deeper ideas and feelings sacred, especially from those you identify as friends and family, and shy away from those who are very opinionated but have no experience in what you are sharing. Those who say they love you may be victims of this parasite, too, and for them, this love is fear-based. Also, keep in mind that many people are programmed to be nosy and newsy. They are automatically compelled to share your private life and ideas, no matter what.

Step two: while you are healing from parasites, let others be who they are. Stop attempting to convince them to ride with you. Allow people to be who they are and stay in your lane. Remember to mind your business and business your mind. When others talk badly about you or don't like you, don't take it personally; they are infected. As Mel Robbins says, you must learn to Let Them be who they are. I suggest you read her book *"Let Them."* It's a good read and a great lesson in becoming free from the parasite of opinion and judgment.

Keep eyes on the road, your hands on the wheel, and never again fall asleep at the wheel on the road from opposition to liberation.

Chapter 22:
America Eats Its Young

In 1972, my favorite band was George Clinton's Funkadelic, later called Parliament-Funkadelic. They released a fantastic album called *"America Eats Its Young"* that I played so much I had to buy a second copy. The cover featured a vampire-ish Statue of Liberty inside US dollar imagery. The album cover was mounted on the wall over my dresser, and the vinyl record stayed on my turntable. The metaphysical messages in their psychedelic-funk music were mind-expanding, insightful, and very deep to me. It influenced my life profoundly.

I'm not totally sure what George Clinton meant by the album title, but here is a bit of my interpretation.

I watched the wounded young soldiers returning from the Vietnam War, literally blown to pieces; some were forced to survive by begging on the same American streets they fought to protect. So many of them were hooked on drugs like heroin and cocaine, attempting to dull the pain; it was sad.

They had been forced to fight a foreign war they didn't understand or necessarily even cared about while dealing with wars in their own neighborhoods at home.

They were being used, eaten, and thrown away by old, greedy, heartless politicians and parasitic banksters. I was 15 and rapidly approaching the draft age. I remember asking the question, "How does this small country called Vietnam threaten the most powerful nation in the world?" One of my relatives who "was served" in the military said it was secretly about massive gold deposits and alien technology, both discovered in Vietnam, that the American vampire republic wanted full control of.

Back at home, people used drugs and food in an attempt to escape the horrors being witnessed on all fronts.

Around this time, food designers were taking advantage of the pain people were feeling by creating food/drugs that alter and eat away the minds, and dulled the senses of the people, especially children, while starving their bodies.

The TV News- lies was involved in hardcore marketing, directing to manipulate minds to go along with this horrible War that we were losing.

They used food as a weapon as the TV was full of commercials convincing a fearful, depressed population to buy cheap, toxic, nutrition-less consumables that technically consumed people and weakened their minds.

Around this time, Howard Moskowitz, a market researcher, sensory science specialist, and psycho physicist, pioneered the concept of the "bliss point," the precise combination of sugar, salt, and fat that makes food and drinks maximally enjoyable and addictive to the human brain, now known as the bliss point.

He perfected the science of how people reacted to flavors, sweetness, and texture. This affects the Nucleus Accumbens, often referred to as "the motivation, learning, and pleasure center" of the brain. He worked with the major food companies, which ultimately made people even more hooked on artificial foods than they already were.

Food manufacturers also marketed these new food-drugs toward children using processed artificial sweeteners like high fructose corn sugar, which should be classified as an additive drug, countless chemicals, poisonous preservatives, and toxic dyes. Using cartoons, slogans, bright colors, and catchy brain entrainment jingles more than ever, programming all those who were unaware of what was happening.

Store owners were convinced to place these items no higher than 3 feet on shelves at children's eye level so that their minds could be snatched. All this programming ultimately took the children of America hostage during the "1970s war of the minds." Once the children are mentally consumed, the parents become powerless and follow suit.

These marketing monsters somehow even got the schools on board with the free toxic lunch programs and placed vending machines in the hallways. These tricks are directed at children, first making parents hostages. Remember the slogan "Tricks are for kids." They were putting the truth right in front of the public's faces.

The evidence of these spiritual wars is all around us today and has changed the neurological systems of most people. A vast population of people now automatically volunteer to join in mental, nutritional, political, and spiritual warfare so that they may see or understand what it is really all about at the root.

Next time I see George Clinton, I will thank him for inspiring me to recognize how many ways America, in fact, does eat its young.

The America spoken of here is not on the physical map; it's a global symbol of a cannibalistic, hypnotic societal menagerie system that turns people into chattel while feeding them full of empty promises and lies, while convincing them to be thankful, happy slaves, while staying prayed up to hopefully get to go to heaven. Did I mention that heaven is not a place; it's a space we can create while living?

The young spoken of here is not just the youth, but people who are mentally like children, still unaware of how a corrupt, greedy system has been designed by parasites to manipulate, confuse, degrade its host while consuming life force energy in perpetuity or until the host wakes up and takes charge of its life.

Growing up, I was called dyslexic, weird, slow, retarded, and crazy. The very traits I was ridiculed for became the gifts I now teach with. A fork-tongued reptilian teacher once told me I was **unteachable**, that I'd be pushing a broom for life, and that I'd never go to college. For many years, I allowed this to consume me until I broke the spell.

I was a college student for exactly 2 ½ weeks. Later in life, after breaking records in the music business, my music production partner and close friend, Tracy Kendrick, and I were asked to teach at UCLA, and we did for four years. UCLA even gave us teaching certificates certifying us to teach anywhere. So, I guess that reptile of a teacher was right, I was unteachable because I was born to be The Teacher.

Parasites have tried to eat me, being almost devoured many times, but for me, every insult became an initiation. Each time, I chose to create rather than conform, as I learned to **"fight the power"**, as the rap group Public Enemy says. Every obstacle has been a test of faith as I realized the biggest enemy is the one within each of us. I have experienced great joys in my life, but also personally suffered extreme sadness, pain, and loss. I still have manic moments, but I realize I am on a mission to be a light in the dark, like the star Sirius, always there to guide us. I love creating Medicine Music, speaking inspiring words, and allowing people to laugh, heal, and learn the true meaning of LOVE.

From Nymph to Dragonfly

The dragonfly life cycle is fascinating to me, and it mirrors themes of this book. A dragonfly begins its life as a nymph living entirely underwater, often in swamps, for up to five years. During this time, it sheds its exoskeleton up to 17 times. These fast underwater predators propel themselves forward with powerful jets of water. They exist in total darkness. They never see sunlight. They never breathe fresh air. They live as both hunters and the hunted, as assailants and victims. Survival, struggle, and fear are all they know, and most don't make it through the first week. Like many beginnings in nature and in our own lives, the early stage is full of chaos, turmoil, and constant opposition.

Then, during the 17th molt, something extraordinary happens. A biological shift begins. Dormant cells inside the nymph awaken and biologically reorganize its entire body. For the first time, it rises out of the water and climbs a reed or twig into the sunlight. There, its hard shell splits open, and the dragonfly emerges. It breathes fresh air for the first time. Its eyes open to a world it has never imagined. It steps out of the old Waterworld of its youth, seeing life from an entirely new perspective.

At this moment, its body is still full of the water it once lived in. Its wings are withered, making flight impossible. Instinctively, it uses the same hydraulic system that powered its underwater life to pump that old swamp water into its wings, expanding and strengthening them. As the sun dries its wings, the dragonfly prepares for its first flight. Then, for the very first time, it lifts off high above the waters of its youth.

For the next 56 days or so, the dragonfly mates every day, creating new life and spreading love everywhere it goes. If you've ever seen two dragonflies mate, they form the shape of a heart, literally expressing love in motion.

Richard Rudd's *The Gene Keys,* a must-read, speaks of the 55th Gene Key, "The Dragonfly Dreams," as a collective shift in human consciousness. My take is this: we are undergoing our own genetic and spiritual mutation. We are rising out of the waters of chaos, turmoil, and opposition just like the dragonfly. But to do this, we must release the paralyzing illusions of separateness, greed, and victimhood. We must pump the swamp water of our past into our wings and allow it to transform us rather than drown us.

Humanity has gone through a long nymph type phase of development. We have lived in survival mode, attacking and consuming each other, trapped in cycles of control, confusion, fear, pain, and fragmentation. We have been hunters and the hunted, assailants and victims. Parasites and hosts. Oppressors and oppressive. Chaos has shaped us, but it has also prepared us to make an evolutionary shift.

Deep within each of us are "imaginal cells," like the caterpillar and the nymph's latent cells, that move us toward metamorphosis. They urge us into transformation, into the next version of ourselves. We are entering a stage of development in which the old parasitic patterns, spiritually, physically, mentally, and energetically, are simply part of the fire of transmutation. They are not here to destroy us; they are here to push us toward our evolution.

We are now rising out of the water.

We are pumping our past into our wings.

We are preparing to fly.

The Alchemist's Revelation

In Paulo Coelho's *The Alchemist*, Santiago spends years searching the world for treasure, only to discover that what he sought was within him all along. His journey mirrors mine. I travel far and endure much, searching for answers, only to return home to the truth that the treasure, the healing, the liberation, the truth has always been inside of me, and now it's in you too. We are the gold we seek. The path itself is the alchemy.

Grandmother's Wisdom

My grandmother, Ethel Peterson, always told me, "Never leave home without a prayer." I've come to understand what she meant. To never leave home without a prayer means to never step into the world without a **direction**, a **goal**, a **telos**, a **destination**. Without it, the opposition will consume you. Prayer, meditation, and being grateful in advance align intention with purpose. It keeps us outcome-oriented and soul-centered.

The Golden Key

Most of the luggage we carry isn't even ours. This book is a spark, a lighthouse, and a signal flare of liberation. May it ignite a fire in you and awaken you from bondage and re-claim your divine power.

You have the keys to break free from the parasite prison of oppression and mental slavery. But only **you** can choose to use these keys.

Like the hero in The Alchemist, you may realize that what you've been seeking has been inside you all along.
It's time to return to your truth, your light, and your divine po-tential.
Be the spell breaker. Reclaim your body, mind, and spirit now, or face the consequences.

As I was coming to the end of this book, the tests came: the computer failed, data was lost, storms came, the winds howled, the rain came blowing sideways, and the windows shook at times. My old saboteur mindset whispered, "Stop now, you're a born loser. You'll never finish this book; you're mentally handicapped."

But I kept writing at times, stopping, starting, breathing deeply, fighting my mental, physical, and energetic parasite demons. The opposition was powerful, but I kept going toward the goal, even though it took seven years.

On 10/22/2025 at 2:22 am, I finally finished this book, broke a spell, and graduated from victim to victory. I celebrated and cried like a baby. If I can do it, you can too.

I have had 13 mentors and guides throughout my life, starting with my parents. They all helped shape me and prepare me for this journey, and I am very grateful. I am here to share all I've learned, inspire you to create a new life, and encourage others to do the same. I have learned that most of the parasites, pirates, oppressors, and their cargo are only opposition there to strengthen us, spar with us, and they show up to test our resolve.

Opposition shows up to be transformed by us so we can go through the resurrection process, be reborn, and rise from the sand like the phoenix.

My Mother, Jessie Branch, has always been my greatest teacher and supporter. She said, "Son, life is like a book; you must keep turning the pages no matter what." Mom, I'm doing just that. I am forever grateful to you.

So now do you have an idea of **What's Eating You**?

Are you ready to take out the trash, lessen the load, and overthrow the parasites on **Your Journey From Opposition To Liberation?**

Travel Light, **Doctah B Sirius**

Elevation Foundation for Abundant Life
ElevationTime.com

Whatseatingyou.life

References

Books

- Virus of The Mind by Richard Dawkins
- Drugs Masquerading as Food by Suzar
- This Is Your Brain on Parasites by Kathleen McAuliffe
- The Science and Romance of Selected Herbs Used in Medicine and Religious Ceremony by Dr. Anthony K. Andoh
- Herbal Secrets of The Rainforest by Leslie Taylor
- Herbs And Things by Jeanne Rose
- Cunningham's Encyclopedia of Magical Herbs by Scott Cunningham
- The Secret Life of Plants by Peter Tompkins and Christopher Bird
- The Hero with a Thousand Faces by Joseph Campbell
- Confessions of a Medical Heretic by Robert S. Mendelson, M.D.
- Guns, Germs, and Steel by Jared Diamond (also a documentary, National Geographic)
- A People's History of the United States by Howard Zinn
- Milk-The Deadly Poison by Robert Cohen
- The Biology of Belief by Dr. Bruce Lipton
- The Four Agreements by Don Miguel Ruiz
- Let Them by Mel Robbins
- The Alchemist by Paulo Coelho
- One Day My Soul Opened Up and Spiritual Hygiene By Iyanla Vanzant

Documentaries, Films, and TV Series

- They Live (1988, John Carpenter)
- The Day the Earth Stood Still (1951)
- The Dark Crystal by Jim Henson
- THE FOOD THAT MADE AMERICA (TV series)
- Cowspiracy

Music

- America Eats Its Young by George Clinton's Funkadelic/Parliament-Funkadelic
- What the world needs now is love not just for one, but for everyone (referenced lyric)

Other Media & References

- The Compendium of the Emerald Tablets by Billy Carson
- The Epic of Humanity by Matthew LaCroix (with Billy Carson)
- CA$HVERTISISING by Drew Whitman

Notable Articles & Reports

- Ethnic Weapons: Race-Specific Biological Weapons, Military Review, November 1970
- FDA Website Statement on Toxoplasma (https://www.fda.gov/Food/ResourcesForYou/HealthEducators/ucm082328.htm)

Notable People Referenced for Their Work

- Masaru Emoto (research on water and words)
- Nevile Goddard (teachings on gratitude and manifestation)
- Sun Tzu (The Art of War, quoted)
- Nikolai Kardashev and Carl Sagan (Type One Civilization, Galactic Federations)

Disclaimer

This book shares natural health and wellness information, historical references, educational insights, traditional practices, and personal research for **informational and educational purposes only**.

The statements made within this publication have **not been evaluated by the U.S. Food and Drug Administration (FDA)**, the American Medical Association (AMA), or any other governmental, medical, or regulatory body. The herbs, tonics, products, and practices mentioned are based on historical usage, cultural traditions, and scientific inquiry, and are intended to support education, awareness, and personal empowerment not to diagnose, treat, cure, or prevent any disease.

We **do not prescribe or provide medical advice**. The information contained herein should never be interpreted as a substitute for professional medical care, diagnosis, or treatment. If you are pregnant, nursing, taking medication, or under a doctor's care, **consult your licensed healthcare provider** before using any natural product or beginning any detoxification or wellness protocol.

By choosing to apply or experiment with any of the educational information, traditional remedies, lifestyle practices, or natural products discussed in this book, **you are exercising your constitutional right to self-care and self-education**. In doing so, you accept full personal responsibility for your own health decisions. Neither the author, publisher, nor any affiliates shall be held liable for any loss, injury, or damage allegedly arising from any information or suggestion contained herein.

As with any food, herb, or natural product, **allergic reactions or sensitivities are possible**. Discontinue use immediately if adverse effects occur.

No statement contained in this book shall be construed as a claim or representation that any product, method, or technique constitutes a diagnosis, cure, mitigation, treatment, or prevention of disease.

While some professionals or researchers may hold differing opinions, the material presented here is offered in good faith as a contribution to ongoing public education in holistic wellness, nutrition, psychology, and energetic science. It reflects both **traditional wisdom and modern scientific curiosity**, and is shared in the spirit of informed choice, self-responsibility, and personal evolution.